2017 SQA Past Papers with Answers

Higher
CHEMISTRY

2015, 2016 & 2017 Exams

HODDER
GIBSON
AN HACHETTE UK COMPANY

This book contains the official 2015, 2016 and 2017 Exams for Higher Chemistry, with associated SQA-approved answers modified from the official marking instructions that accompany the paper.

In addition the book contains study skills advice. This advice has been specially commissioned by Hodder Gibson, and has been written by experienced senior teachers and examiners in line with the Higher for CfE syllabus and assessment outlines. This is not SQA material but has been devised to provide further guidance for Higher examinations.

Hodder Gibson is grateful to the copyright holders, as credited on the final page of the Answer section, for permission to use their material. Every effort has been made to trace the copyright holders and to obtain their permission for the use of copyright material. Hodder Gibson will be happy to receive information allowing us to rectify any error or omission in future editions.

Hachette UK's policy is to use papers that are natural, renewable and recyclable products and made from wood grown in sustainable forests. The logging and manufacturing processes are expected to conform to the environmental regulations of the country of origin.

Orders: please contact Bookpoint Ltd, 130 Park Drive, Milton Park, Abingdon, Oxon OX14 4SE. Telephone: (44) 01235 827720. Fax: (44) 01235 400454. Lines are open 9.00–5.00, Monday to Saturday, with a 24-hour message answering service. Visit our website at www.hoddereducation.co.uk. Hodder Gibson can be contacted direct on: Tel: 0141 333 4650; Fax: 0141 404 8188; email: hoddergibson@hodder.co.uk

This collection first published in 2017 by
Hodder Gibson, an imprint of Hodder Education,
An Hachette UK Company
211 St Vincent Street
Glasgow G2 5QY

Typeset by Aptara, Inc.

Printed in the UK

A catalogue record for this title is available from the British Library

ISBN: 978-1-5104-2144-8

2 1

2018 2017

Introduction

Study Skills – what you need to know to pass exams!

Pause for thought

Many students might skip quickly through a page like this. After all, we all know how to revise. Do you really though?

Think about this:

"IF YOU ALWAYS DO WHAT YOU ALWAYS DO, YOU WILL ALWAYS GET WHAT YOU HAVE ALWAYS GOT."

Do you like the grades you get? Do you want to do better? If you get full marks in your assessment, then that's great! Change nothing! This section is just to help you get that little bit better than you already are.

There are two main parts to the advice on offer here. The first part highlights fairly obvious things but which are also very important. The second part makes suggestions about revision that you might not have thought about but which WILL help you.

Part 1

DOH! It's so obvious but …

Start revising in good time

Don't leave it until the last minute – this will make you panic.

Make a revision timetable that sets out work time AND play time.

Sleep and eat!

Obvious really, and very helpful. Avoid arguments or stressful things too – even games that wind you up. You need to be fit, awake and focused!

Know your place!

Make sure you know exactly **WHEN and WHERE** your exams are.

Know your enemy!

Make sure you know what to expect in the exam.

How is the paper structured?

How much time is there for each question?

What types of question are involved?

Which topics seem to come up time and time again?

Which topics are your strongest and which are your weakest?

Are all topics compulsory or are there choices?

Learn by DOING!

There is no substitute for past papers and practice papers – they are simply essential! Tackling this collection of papers and answers is exactly the right thing to be doing as your exams approach.

Part 2

People learn in different ways. Some like low light, some bright. Some like early morning, some like evening / night. Some prefer warm, some prefer cold. But everyone uses their BRAIN and the brain works when it is active. Passive learning – sitting gazing at notes – is the most INEFFICIENT way to learn anything. Below you will find tips and ideas for making your revision more effective and maybe even more enjoyable. What follows gets your brain active, and active learning works!

Activity 1 – Stop and review

Step 1

When you have done no more than 5 minutes of revision reading STOP!

Step 2

Write a heading in your own words which sums up the topic you have been revising.

Step 3

Write a summary of what you have revised in no more than two sentences. Don't fool yourself by saying, "I know it, but I cannot put it into words". That just means you don't know it well enough. If you cannot write your summary, revise that section again, knowing that you must write a summary at the end of it. Many of you will have notebooks full of blue/black ink writing. Many of the pages will not be especially attractive or memorable so try to liven them up a bit with colour as you are reviewing and rewriting. **This is a great memory aid, and memory is the most important thing.**

Activity 2 – Use technology!

Why should everything be written down? Have you thought about "mental" maps, diagrams, cartoons and colour to help you learn? And rather than write down notes, why not record your revision material?

What about having a text message revision session with friends? Keep in touch with them to find out how and what they are revising and share ideas and questions.

Why not make a video diary where you tell the camera what you are doing, what you think you have learned and what you still have to do? No one has to see or hear it, but the process of having to organise your thoughts in a formal way to explain something is a very important learning practice.

Be sure to make use of electronic files. You could begin to summarise your class notes. Your typing might be slow, but it will get faster and the typed notes will be easier to read than the scribbles in your class notes. Try to add different fonts and colours to make your work stand out. You can easily Google relevant pictures, cartoons and diagrams which you can copy and paste to make your work more attractive and **MEMORABLE**.

Activity 3 – This is it. Do this and you will know lots!

Step 1

In this task you must be very honest with yourself! Find the SQA syllabus for your subject (www.sqa.org.uk). Look at how it is broken down into main topics called MANDATORY knowledge. That means stuff you MUST know.

Step 2

BEFORE you do ANY revision on this topic, write a list of everything that you already know about the subject. It might be quite a long list but you only need to write it once. It shows you all the information that is already in your long-term memory so you know what parts you do not need to revise!

Step 3

Pick a chapter or section from your book or revision notes. Choose a fairly large section or a whole chapter to get the most out of this activity.

With a buddy, use Skype, Facetime, Twitter or any other communication you have, to play the game "If this is the answer, what is the question?". For example, if you are revising Geography and the answer you provide is "meander", your buddy would have to make up a question like "What is the word that describes a feature of a river where it flows slowly and bends often from side to side?".

Make up 10 "answers" based on the content of the chapter or section you are using. Give this to your buddy to solve while you solve theirs.

Step 4

Construct a wordsearch of at least 10 × 10 squares. You can make it as big as you like but keep it realistic. Work together with a group of friends. Many apps allow you to make wordsearch puzzles online. The words and phrases can go in any direction and phrases can be split. Your puzzle must only contain facts linked to the topic you are revising. Your task is to find 10 bits of information to hide in your puzzle, but you must not repeat information that you used in Step 3. DO NOT show where the words are. Fill up empty squares with random letters. Remember to keep a note of where your answers are hidden but do not show your friends. When you have a complete puzzle, exchange it with a friend to solve each other's puzzle.

Step 5

Now make up 10 questions (not "answers" this time) based on the same chapter used in the previous two tasks. Again, you must find NEW information that you have not yet used. Now it's getting hard to find that new information! Again, give your questions to a friend to answer.

Step 6

As you have been doing the puzzles, your brain has been actively searching for new information. Now write a NEW LIST that contains only the new information you have discovered when doing the puzzles. Your new list is the one to look at repeatedly for short bursts over the next few days. Try to remember more and more of it without looking at it. After a few days, you should be able to add words from your second list to your first list as you increase the information in your long-term memory.

FINALLY! Be inspired...

Make a list of different revision ideas and beside each one write **THINGS I HAVE** tried, **THINGS I WILL** try and **THINGS I MIGHT** try. Don't be scared of trying something new.

And remember – "FAIL TO PREPARE AND PREPARE TO FAIL!"

Higher Chemistry

The Course

The main aims of the Higher Chemistry course are for learners to:

- develop and apply knowledge and understanding of chemistry

- develop an understanding of chemistry's role in scientific issues and relevant applications of chemistry, including the impact these could make in society and the environment

- develop scientific analytical thinking skills, including scientific evaluation, in a chemistry context

- develop the use of technology, equipment and materials, safely, in practical scientific activities, including using risk assessments

- develop scientific inquiry, investigative, problem solving and planning skills

- use and understand scientific literacy to communicate ideas and issues and to make scientifically informed choices

- develop skills of independent working.

How the Course is assessed

To gain the Course award:

(i) you must pass the four units – Chemical Changes and Structure, Researching Chemistry, Nature's Chemistry and Chemistry in Society. The units are assessed internally on a pass/fail basis.

(ii) you must submit an assignment which is externally marked by the SQA and is worth 20 marks.

(iii) you must sit the Higher Chemistry exam paper which is set and marked by the SQA and is worth 100 marks.

The course award is graded A–D, the grade being determined by the total mark you score in the examination and the mark you gain in the assignment.

The Examination

- The examination consists of one exam paper which has two sections and lasts 2 hours 30 minutes:

  Section 1 (multiple choice)   20 marks

  Section 2 (extended answer)   80 marks

Further details can be found in the Higher Chemistry section on the SQA website:

http://www.sqa.org.uk/sqa/47913.html

Key Tips For Your Success

Practise! Practise! Practise!

In common with Higher Mathematics and the other Higher sciences, the key to exam success in Chemistry is to prepare by regularly answering questions. Use the questions as a prompt for further study: if you find that you cannot answer a question, review your notes and/or textbook to help you find the necessary knowledge to answer the question. You will quickly find out what you can/cannot do if you invest time attempting to answer questions. It is a much more valuable use of time than passively copying notes, which is a common trap many students fall into!

The data booklet

The data booklet contains formulae and useful data, which you will have to use in the exam. Although you might think that you have a good memory for chemical data (such as the symbols for elements or the atomic mass of an element) always check using the data booklet.

Calculations

In preparation for the exam, ensure that you recognise the different calculation types:

- relative rate

- using bond enthalpy

- using $cm\Delta T$

- percentage yield

- atom economy

- using molar volume

- volumetric calculations

- calculations from balanced chemical equations

You will encounter these calculations in the exam so it's worth spending time practising to ensure that you are familiar with the routines for solving these problems. Even if you are not sure how to attempt a calculation question, show your working! You will be given credit for calculations, which are relevant to the problem being solved.

Explain questions

You will encounter questions, which ask you to *explain your answer*. Take your time and attempt to explain to the examiner. If you can use a diagram or chemical equations to aid your answer, use these as they can really bring an answer to life.

Applying your knowledge of practical chemistry

As part of your Higher Chemistry experience, you should have had plenty of practice carrying out experiments using standard lab equipment and you should have had opportunities to evaluate your results from experiments. In the Higher exam, you are expected to be familiar with the techniques and apparatus listed in the tables below.

Apparatus

Beaker	Dropper	Pipette filler
Boiling tube	Evaporating basin	Test tubes
Burette	Funnel	Thermometer
Conical flask	Measuring cylinder	Volumetric flask
Delivery tubes	Pipette	

Techniques

Distillation
Filtration
Methods for collecting a gas: over water or using a gas syringe
Safe heating methods: using a Bunsen, water bath or heating mantle
Titration
Use of a balance

The following general points about experimental chemistry are worth noting:

- A pipette is more accurate than a measuring cylinder for measuring fixed volumes of liquid. A burette can be used to measure non-standard volumes of liquid.

- A standard flask is used to make up a standard solution i.e. a solution of accurately known concentration. This is done by dissolving a known mass of solute in water and transferring to the standard flask with rinsings. Finally, the standard flask is made up to the mark with water.

- A gas syringe is an excellent method for measuring the volume of gas produced from an experiment.

- Bunsen burners cannot be used near flammable substances.

- A Bunsen burner does not allow you to control the rate of heating.

Analysis of data

From your experience working with experimental data you should know how to calculate averages, how to eliminate rogue data, how to draw graphs (scatter and best fit line/curve) and how to interpret graphs.

It is common in Higher exams to be presented with titration data such as the data shown in the table below.

Titration	Volume of solution cm^3
1	26.0
2	24.1
3	39.0
4	24.2
5	24.8

Result 1 is a rough titration which is not accurate.

Results 2 and 4 could be used to calculate an average volume (= $24.15cm^3$)

Result 3 is a rogue result and should be ignored.

Result 5 cannot be used to calculate the average volume as it is too far from 24.1 and 24.2 i.e. it is not accurate.

Numeracy

The Higher Chemistry exam will contain several questions that test your numeracy skills e.g. calculating relative rate, enthalpy changes, percentage yield etc. Other questions will ask you to "scale up" or "scale down" as this is a skill that is used by practising scientists in their day to day job.

Being able to deal with proportion is key to answering numeracy questions in chemistry. A common layout is shown in the examples below. In all cases, the unknown (what you are being asked to calculate) should be put on the right hand side.

[Example]

1.2g of methane burned to produce 52kJ of energy. Calculate the enthalpy of combustion of methane.

Answer:

This is really a proportion question. You have to understand that the enthalpy of combustion is the energy released when 1 mole of a substance is burned completely, and that 1 mole of methane is the gram formula mass i.e. 16g.

Step 1: State a relationship

Mass Energy

1.2g ➡ 52kJ

(note that energy is placed on the right hand side as we want to calculate the energy)

Step 2: Scale to 1

1g ➡ = 43.3kJ

Step 3: Calculate for the mass you are asked

16g ➡ 16x 43.3= 693.3

Finally, answer the question: The ΔH combustion for methane = -693.3kJmol^{-1}

[Example]

A 100ml bottle of children's paracetamol costs £3.85. The ingredients label states that each 5ml dose contains 120mg of paracetamol. Calculate the cost per mg of paracetamol.

Answer:

Volume		Mass
5ml	➡	120mg
1ml	➡	24mg
100ml	➡	2400mg

i.e. 1 bottle contains 2400mg of paracetamol

Mass		Cost
2400mg	➡	£3.85
1mg	➡	£0.0016

Open-ended questions

Real-life chemistry problems rarely have a fixed answer. In the Higher exam, you will encounter two 3 mark questions that are open-ended i.e. there is more than one "correct" answer. You will recognise these questions from the phrase *using your knowledge of chemistry* in the question. To tackle these, look at the following example.

[Example]

SQA Higher Chemistry 2013 Q.12

Cooking involves many chemical reactions. Proteins, fats, oils and esters are some examples of compounds found in food. A chemist suggested that cooking food could change compounds from being fat-soluble to water-soluble.

Use your knowledge of chemistry to comment on the accuracy of this statement.

Author's suggested answer

To tackle a question like this, focus on the key chemical words and think about the chemistry you know. What chemical reactions do you know that involve proteins, fats, oils and esters? Can you relate this to solubility?

Proteins - Long chain molecules linked by hydrogen bonding. Perhaps the proteins in food are insoluble as the chains are attracted to themselves. Cooking could cause the protein chains to untwist (breaking the hydrogen bonds) making them more likely to attract water to the exposed peptide links. In addition, cooking could cause the protein to hydrolyse to produce amino acids. Amino acids contain the polar amine group (–NH$_2$) and carboxyl group (–COOH) – both can form hydrogen bonds to water, therefore the amino acids can dissolve in water.

Fats and Oils - Insoluble in water as they are mainly large hydrocarbon structures. Fats and oils can hydrolyse to produce glycerol and fatty acids. Glycerol has three –OH groups so it could H-bond to water molecules and dissolve. Fatty acids contain a polar head (the carboxyl group-COOH) which is water soluble.

Esters - Non-polar and insoluble. Heating could hydrolyse the ester group producing an alcohol and carboxylic acid. Both of these molecules are polar and would dissolve in water.

A good answer for this question wouldn't have to contain all of the above. Indeed, it could focus on one molecule but give lots of detail. It's also a good idea to illustrate your answer with diagrams e.g. you could show typical structures and show how they can bond to water. If it enhances your answer by showing the examiner that you understand the chemistry, include it!

Good luck!

If you have followed the advice given in this introduction you will be well prepared for the Higher exam. When you sit the exam, take your time and use the experience as an opportunity to show the examiner how much you know. And good luck!

HIGHER

2015

National Qualifications 2015

X713/76/02

Chemistry
Section 1 — Questions

THURSDAY, 28 MAY

1:00 PM – 3:30 PM

Instructions for the completion of Section 1 are given on *Page two* of your question and answer booklet X713/76/01.

Record your answers on the answer grid on *Page three* of your question and answer booklet.

Reference may be made to the Chemistry Higher and Advanced Higher Data Booklet.

Before leaving the examination room you must give your question and answer booklet to the Invigilator; if you do not, you may lose all the marks for this paper.

SECTION 1 — 20 marks
Attempt ALL questions

1. The elements nitrogen, oxygen, fluorine and neon

   A   can form negative ions

   B   are made up of diatomic molecules

   C   have single bonds between the atoms

   D   are gases at room temperature.

2. Which of the following equations represents the first ionisation energy of fluorine?

   A $F^-(g) \rightarrow F(g) + e^-$

   B $F^-(g) \rightarrow \frac{1}{2}F_2(g) + e^-$

   C $F(g) \rightarrow F^+(g) + e^-$

   D $\frac{1}{2}F_2(g) \rightarrow F^+(g) + e^-$

3. Which of the following atoms has least attraction for bonding electrons?

   A   Carbon

   B   Nitrogen

   C   Phosphorus

   D   Silicon

4. Which of the following is **not** an example of a van der Waals' force?

   A   Covalent bond

   B   Hydrogen bond

   C   London dispersion force

   D   Permanent dipole – permanent dipole attraction

5. Which of the following has more than one type of van der Waals' force operating between its molecules in the liquid state?

   A   Br—Br

   B   O=C=O

   C   H–N with H, H (ammonia structure)

   D   H–C with H, H, H (methane structure)

6. Oil molecules are more likely to react with oxygen in the air than fat molecules.

   During the reaction the oil molecules

   A   are reduced

   B   become rancid

   C   are hydrolysed

   D   become unsaturated.

7. Which of the following mixtures will form when NaOH(aq) is added to a mixture of propanol and ethanoic acid?

   A   Propanol and sodium ethanoate

   B   Ethanoic acid and sodium propanoate

   C   Sodium hydroxide and propyl ethanoate

   D   Sodium hydroxide and ethyl propanoate

8. Oils contain carbon to carbon double bonds which can undergo addition reactions with iodine.

   The iodine number of an oil is the mass of iodine in grams that will react with 100 g of oil.

   Which line in the table shows the oil that is likely to have the lowest melting point?

	Oil	Iodine number
A	Corn	123
B	Linseed	179
C	Olive	81
D	Soya	130

[Turn over

9. When an oil is hydrolysed, which of the following molecules is always produced?

A
$$
\begin{array}{c}
COOH \\
| \\
CHOH \\
| \\
COOH
\end{array}
$$

B
$$
\begin{array}{c}
CH_2OH \\
| \\
CHOH \\
| \\
CH_2OH
\end{array}
$$

C $C_{17}H_{35}COOH$

D $C_{17}H_{33}COOH$

10. Enzymes are involved in the browning of cut fruit.

One reaction taking place is:

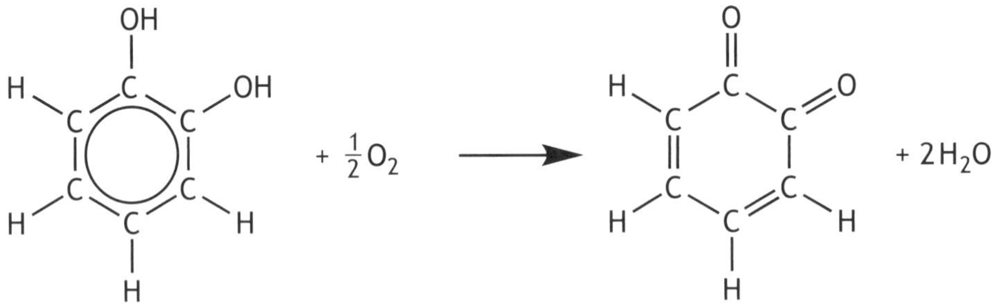

Which of the following correctly describes the above reaction?

A Oxidation

B Reduction

C Hydrolysis

D Condensation

11. Which of the following statements is correct for ketones?

A They are formed by oxidation of tertiary alcohols.

B They contain the group 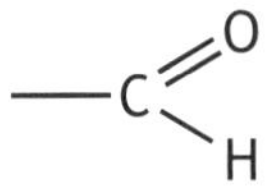 .

C They contain a carboxyl group.

D They will not react with Fehling's solution.

12. Carvone is a natural product that can be extracted from orange peel.

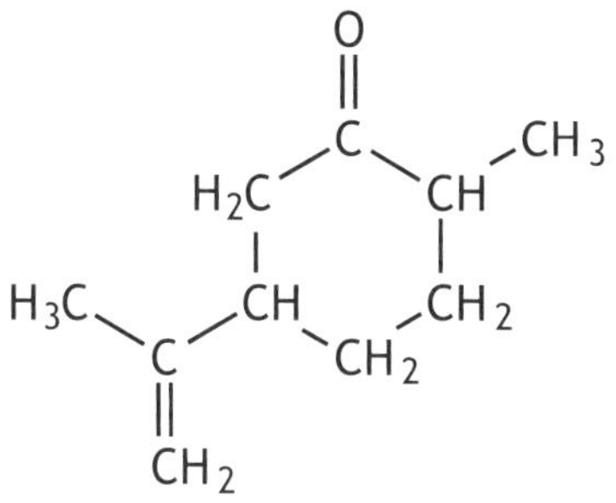

Carvone

Which line in the table correctly describes the reaction of carvone with bromine solution and with acidified potassium dichromate solution?

	Reaction with bromine solution	Reaction with acidified potassium dichromate solution
A	no reaction	no reaction
B	no reaction	orange to green
C	decolourises	orange to green
D	decolourises	no reaction

13. The structure of isoprene is

A

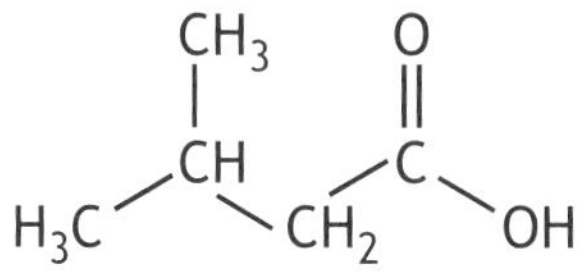

B

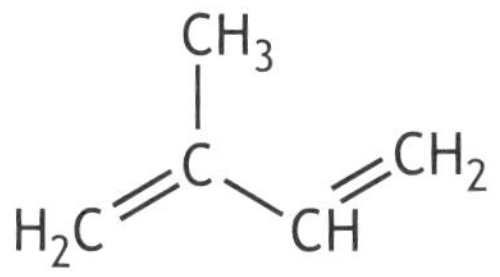

C

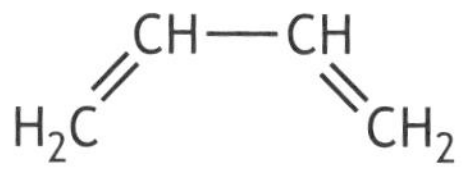

D

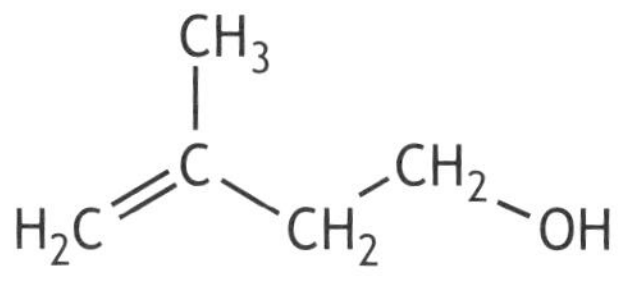

[Turn over

14. The antibiotic, erythromycin, has the following structure.

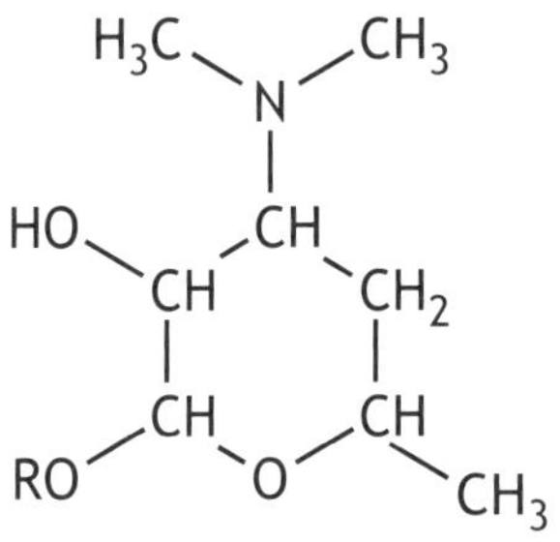

To remove its bitter taste, the erythromycin is reacted to give the compound with the structure shown below.

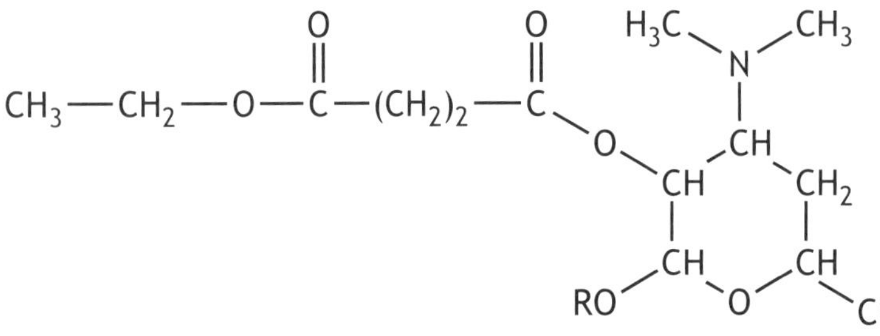

Which of the following types of compound has been reacted with erythromycin to produce this compound?

A Alcohol

B Aldehyde

C Carboxylic acid

D Ketone

15. Which of the following is an isomer of 2,2-dimethylpentan-1-ol?

A $CH_3CH_2CH_2CH(CH_3)CH_2OH$

B $(CH_3)_3CCH(CH_3)CH_2OH$

C $CH_3CH_2CH_2CH_2CH_2CH_2CH_2CH_2OH$

D $(CH_3)_2CHC(CH_3)_2CH_2CH_2OH$

16. Consider the reaction pathway shown below.

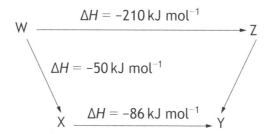

According to Hess's Law, the ΔH value, in kJ mol^{-1}, for reaction Z to Y is

A   +74

B   −74

C   +346

D   −346.

[Turn over

17. $I_2(s) \rightarrow I_2(g)$ $\Delta H = +60 \, kJ \, mol^{-1}$

$I_2(g) \rightarrow 2I(g)$ $\Delta H = +243 \, kJ \, mol^{-1}$

$I(g) + e^- \rightarrow I^-(g)$ $\Delta H = -349 \, kJ \, mol^{-1}$

Which of the following would show the energy diagram for $I_2(s) + 2e^- \rightarrow 2I^-(g)$?

A

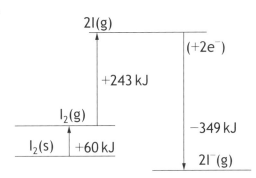

B

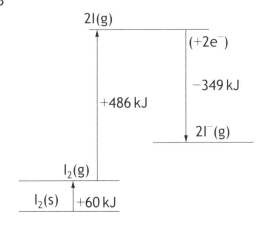

C

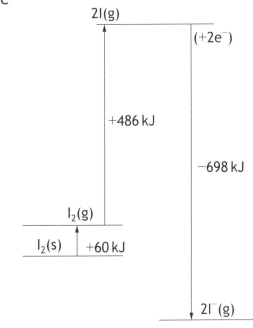

D

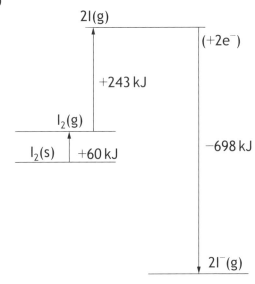

18. Which of the following statements regarding a chemical reaction at equilibrium is always correct?

    A   The rates of the forward and reverse reactions are equal.

    B   The concentration of reactants and products are equal.

    C   The forward and reverse reactions have stopped.

    D   The addition of a catalyst changes the position of the equilibrium.

19. A reaction has the following potential energy diagram.

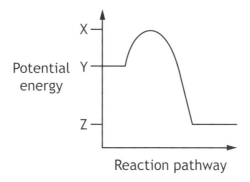

    The activation energy for the forward reaction is

    A   X – Y

    B   Y – X

    C   Y – Z

    D   Z – Y.

20. Which of the following will react with Br_2 but **not** with I_2?

    A OH^-

    B SO_3^{2-}

    C Fe^{2+}

    D Mn^{2+}

**[END OF SECTION 1. NOW ATTEMPT THE QUESTIONS IN SECTION 2
OF YOUR QUESTION AND ANSWER BOOKLET.]**

[BLANK PAGE]

DO NOT WRITE ON THIS PAGE

H

National
Qualifications
2015

Mark

X713/76/01

**Chemistry
Section 1 — Answer Grid
and Section 2**

THURSDAY, 28 MAY

1:00 PM – 3:30 PM

Fill in these boxes and read what is printed below.

Full name of centre

Town

Forename(s)

Surname

Number of seat

Date of birth

| Day | Month | Year |

Scottish candidate number

Total marks — 100

SECTION 1 — 20 marks

Attempt ALL questions.

Instructions for completion of Section 1 are given on *Page two*.

SECTION 2 — 80 marks

Attempt ALL questions

Reference may be made to the Chemistry Higher and Advanced Higher Data Booklet.

Write your answers clearly in the spaces provided in this booklet. Additional space for answers and rough work is provided at the end of this booklet. If you use this space you must clearly identify the question number you are attempting. Any rough work must be written in this booklet. You should score through your rough work when you have written your final copy.

Use **blue** or **black** ink.

Before leaving the examination room you must give this booklet to the Invigilator; if you do not, you may lose all the marks for this paper.

SECTION 1— 20 marks

The questions for Section 1 are contained in the question paper X713/76/02.
Read these and record your answers on the answer grid on *Page three* opposite.
Use **blue** or **black** ink. Do NOT use gel pens or pencil.

1. The answer to each question is **either** A, B, C or D. Decide what your answer is, then fill in the appropriate bubble (see sample question below).

2. There is **only one correct** answer to each question.

3. Any rough working should be done on the additional space for answers and rough work at the end of this booklet.

Sample Question

To show that the ink in a ball-pen consists of a mixture of dyes, the method of separation would be:

    A   fractional distillation

    B   chromatography

    C   fractional crystallisation

    D   filtration.

The correct answer is **B**—chromatography. The answer **B** bubble has been clearly filled in (see below).

Changing an answer

If you decide to change your answer, cancel your first answer by putting a cross through it (see below) and fill in the answer you want. The answer below has been changed to **D**.

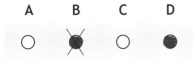

If you then decide to change back to an answer you have already scored out, put a tick (✓) to the **right** of the answer you want, as shown below:

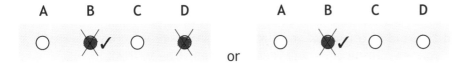

SECTION 1 — Answer Grid

	A	B	C	D
1	○	○	○	○
2	○	○	○	○
3	○	○	○	○
4	○	○	○	○
5	○	○	○	○
6	○	○	○	○
7	○	○	○	○
8	○	○	○	○
9	○	○	○	○
10	○	○	○	○
11	○	○	○	○
12	○	○	○	○
13	○	○	○	○
14	○	○	○	○
15	○	○	○	○
16	○	○	○	○
17	○	○	○	○
18	○	○	○	○
19	○	○	○	○
20	○	○	○	○

MARKS |

SECTION 2 — 80 marks

Attempt ALL questions

1. Volcanoes produce a variety of molten substances, including sulfur and silicon dioxide.

   (a) Complete the table to show the strongest type of attraction that is broken when each substance melts.

Substance	Melting point (°C)	Strongest type of attraction broken when substance melts
sulfur	113	
silicon dioxide	1610	

   2

   (b) Volcanic sulfur can be put to a variety of uses. One such use involves reacting sulfur with phosphorus to make a compound with formula P_4S_3.

      (i) Draw a possible structure for P_4S_3.

      1

      (ii) Explain why the covalent radius of sulfur is smaller than that of phosphorus.

      1

MARKS | DO NOT WRITE IN THIS MARGIN

1. (b) (continued)

(iii) The melting point of sulfur is much higher than that of phosphorus.

Explain fully, in terms of the structures of sulfur and phosphorus molecules and the intermolecular forces between molecules of each element, why the melting point of sulfur is much higher than that of phosphorus.

3

[Turn over

2. (a) A student investigated the effect of changing acid concentration on reaction rate. Identical strips of magnesium ribbon were dropped into different concentrations of excess hydrochloric acid and the time taken for the magnesium to completely react recorded.

A graph of the student's results is shown below.

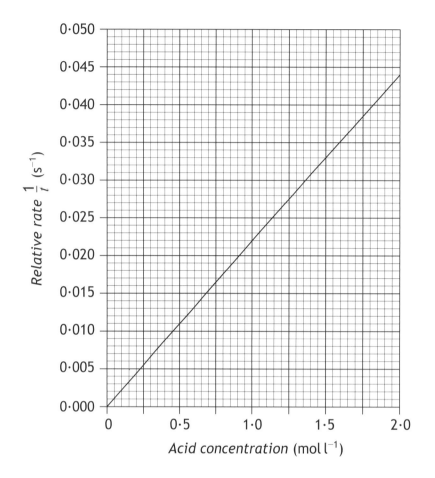

Use information from the graph to calculate the reaction time, in seconds, when the concentration of the acid was $1 \cdot 0$ mol l^{-1}.   **1**

MARKS | DO NOT WRITE IN THIS MARGIN

2. (continued)

(b) The rate of reaction can also be altered by changing the temperature or using a catalyst.

(i) Graph 1 shows the distribution of kinetic energies of molecules in a gas at 100 °C.

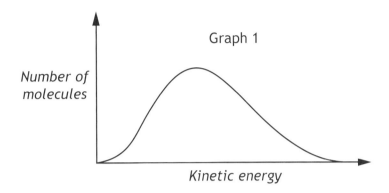

Add a second curve to graph 1 to show the distribution of kinetic energies at 50 °C.

1

(ii) In graph 2, the shaded area represents the number of molecules with the required activation energy, E_a.

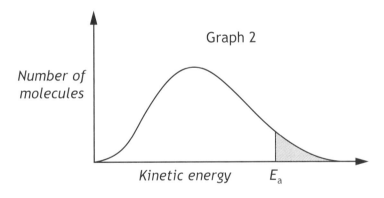

Draw a line to show how a catalyst affects the activation energy.

1

[Turn over

MARKS | DO NOT WRITE IN THIS MARGIN

3. (a) Methyl cinnamate is an ester used to add strawberry flavour to foods. It is a naturally occurring ester found in the essential oil extracted from the leaves of strawberry gum trees.

To extract the essential oil, steam is passed through shredded strawberry gum leaves. The steam and essential oil are then condensed and collected.

(i) Complete the diagram to show an apparatus suitable for carrying out this extraction.

(An additional diagram, if required, can be found on *Page thirty-five*).

2

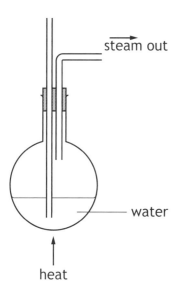

steam out

water

heat

(ii) The essential oil extracted is a mixture of compounds.

Suggest a technique that could be used to separate the mixture into pure compounds.

1

(b) A student prepared a sample of methyl cinnamate from cinnamic acid and methanol.

cinnamic acid   +   methanol   →   methyl cinnamate   +   water

mass of one mole   mass of one mole   mass of one mole
    = 148 g   = 32 g   = 162 g

6·5 g of cinnamic acid was reacted with 2·0 g of methanol.

MARKS | DO NOT WRITE IN THIS MARGIN

3. (b) (continued)

(i) Show, by calculation, that cinnamic acid is the limiting reactant. (One mole of cinnamic acid reacts with one mole of methanol.)   2

(ii) (A) The student obtained 3·7 g of methyl cinnamate from 6·5 g of cinnamic acid.

Calculate the percentage yield.   2

(B) The student wanted to scale up the experiment to make 100 g of methyl cinnamate.

Cinnamic acid costs £35·00 per 250 g.

Calculate the cost of cinnamic acid needed to produce 100 g of methyl cinnamate.   2

[Turn over

4. Up to 10% of perfumes sold in the UK are counterfeit versions of brand name perfumes.

One way to identify if a perfume is counterfeit is to use gas chromatography. Shown below are gas chromatograms from a brand name perfume and two different counterfeit perfumes. Some of the peaks in the brand name perfume have been identified as belonging to particular compounds.

Brand name perfume

(A) linalool

(B) citronellol

(C) geraniol

(D) eugenol

(E) anisyl alcohol

(F) coumarin

(G) benzyl salicylate

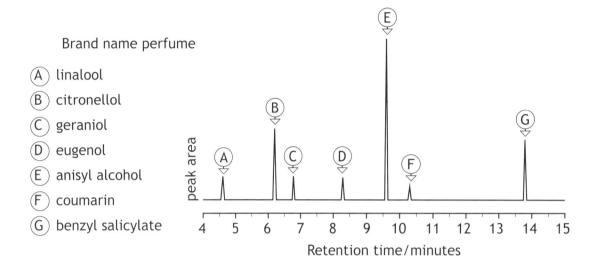

Counterfeit A

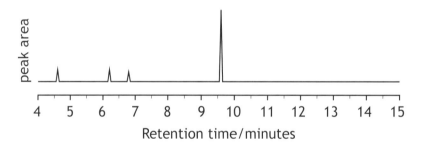

Counterfeit B

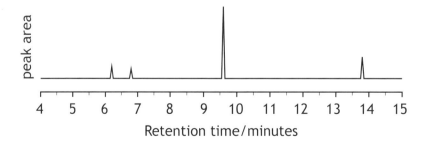

MARKS | DO NOT WRITE IN THIS MARGIN

4. (continued)

(a) Identify one compound present in the brand name perfume that appears in both counterfeit perfumes.

1

(b) Some compounds in the brand name perfume are not found in the counterfeit perfumes. State another difference that the chromatograms show between the counterfeit perfumes and the brand name perfume.

1

(c) The gas used to carry the perfume sample along the chromatography column is helium.

(i) Suggest why helium is used.

1

(ii) Apart from the polarity of the molecules, state another factor that would affect the retention time of molecules during gas chromatography.

1

[Turn over

MARKS | DO NOT WRITE IN THIS MARGIN

4.   (continued)

(d) Many of the compounds in perfumes are molecules consisting of joined isoprene units.

(i) State the name that is given to molecules consisting of joined isoprene units.   **1**

(ii) Geraniol is one of the compounds found in perfume. It has the following structural formula and systematic name.

3,7-dimethylocta-2,6-dien-1-ol

Linalool can also be present. Its structural formula is shown.

(A)   State the systematic name for linalool.   **1**

(B)   Explain why linalool can be classified as a tertiary alcohol.   **1**

MARKS | DO NOT WRITE IN THIS MARGIN

4. (continued)

(e) Coumarin is another compound found in the brand name perfume. It is present in the spice cinnamon and can be harmful if eaten in large quantities.

The European Food Safety Authority gives a tolerable daily intake of coumarin at 0·10 mg per kilogram of body weight.

1·0 kg of cinnamon powder from a particular source contains 4·4 g of coumarin. Calculate the mass of cinnamon powder, in g, which would need to be consumed by an adult weighing 75 kg to reach the tolerable daily intake.

2

[Turn over

5. **Patterns in the Periodic Table**

The Periodic Table is an arrangement of all the known elements in order of increasing atomic number. The reason why the elements are arranged as they are in the Periodic Table is to fit them all, with their widely diverse physical and chemical properties, into a logical pattern.

Periodicity is the name given to regularly-occurring similarities in physical and chemical properties of the elements.

Some Groups exhibit striking similarity between their elements, such as Group 1, and in other Groups the elements are less similar to each other, such as Group 4, but each Group has a common set of characteristics.

Adapted from Royal Society of Chemistry, Visual Elements (rsc.org)

Using your knowledge of chemistry, comment on similarities and differences in the patterns of physical and chemical properties of elements in both Group 1 and Group 4.

3

MARKS | DO NOT WRITE IN THIS MARGIN

6. Uncooked egg white is mainly composed of dissolved proteins. During cooking processes, the proteins become denatured as the protein chains unwind, and the egg white solidifies.

(a) Explain why the protein chains unwind. **1**

(b) The temperature at which the protein becomes denatured is called the melting temperature. The melting temperature of a protein can be determined using fluorescence. In this technique, the protein is mixed with a dye that gives out visible light when it attaches to hydrophobic parts of the protein molecule. The hydrophobic parts of the structure are on the inside of the protein and the dye has no access to them unless the protein unwinds.

(i) Ovalbumin is a protein found in egg white. Part of the structure of unwound ovalbumin is shown below.

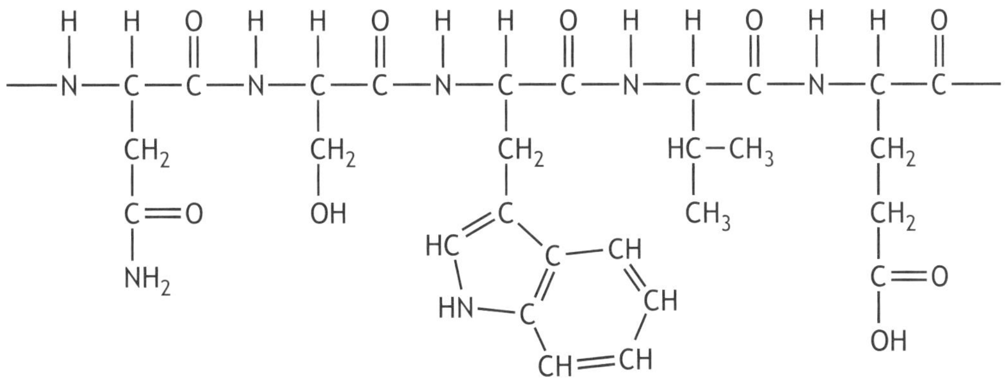

Circle the part of the structure to which the hydrophobic dye is most likely to attach. **1**

[Turn over

MARKS | DO NOT WRITE IN THIS MARGIN

6. (b) (continued)

   (ii) Another protein in egg white is conalbumin. The temperature of a conalbumin/dye mixture is gradually increased. The fluorescence is measured and a graph is produced.

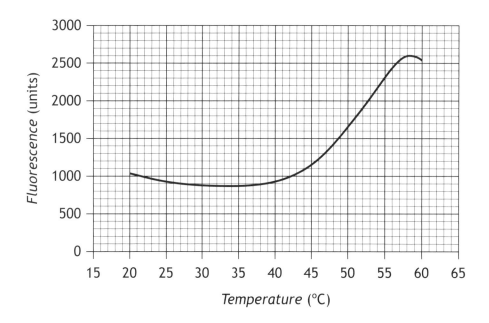

The melting temperature is the temperature at which the fluorescence is halfway between the highest and lowest fluorescence values.

Determine the melting temperature, in °C, for this protein. 1

MARKS | DO NOT WRITE IN THIS MARGIN

6. (continued)

(c) Once cooked and eaten, the digestive system breaks the protein chains into amino acids with the help of enzymes.

  (i) State the name of the digestion process where enzymes break down proteins into amino acids.   **1**

  (ii)

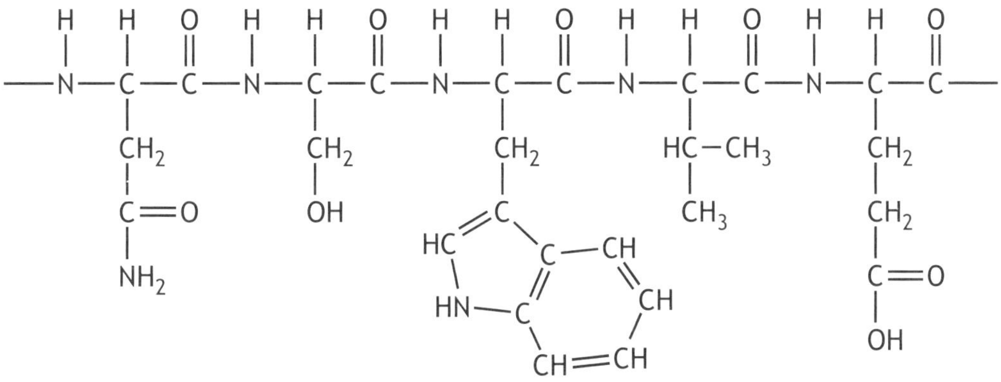

  (A) State how many amino acid molecules joined to form this section of protein.   **1**

  (B) Draw the structure of one amino acid that would be produced when this section of the protein chain is broken down.   **1**

MARKS | DO NOT WRITE IN THIS MARGIN

7. Methanol can be used as a fuel, in a variety of different ways.

$$H-\underset{\underset{H}{|}}{\overset{\overset{H}{|}}{C}}-OH$$

(a) An increasingly common use for methanol is as an additive in petrol.

Methanol has been tested as an additive in petrol at 118 g per litre of fuel.

Calculate the volume of carbon dioxide, in litres, that would be released by combustion of 118 g of methanol.

$$2CH_3OH(\ell) \quad + \quad 3O_2(g) \quad \rightarrow \quad 2CO_2(g) \quad + \quad 4H_2O(\ell)$$

(Take the molar volume of carbon dioxide to be 24 litres mol^{-1}.)   **2**

MARKS | DO NOT WRITE IN THIS MARGIN

7. (continued)

    (b) A student investigated the properties of methanol and ethanol.

        (i) The student carried out experiments to determine the enthalpy of combustion of the alcohols.

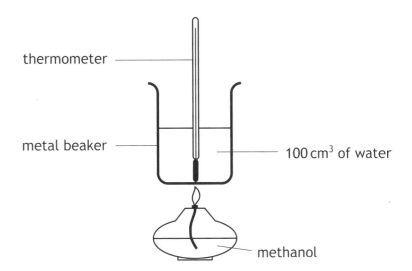

        (A) The student carried out the first experiment as shown, but was told to repeat the experiment as the thermometer had been placed in the wrong position.

            Suggest why the student's placing of the thermometer was incorrect.   1

        (B) The student always used 100 cm³ of water.

            State another variable that the student should have kept constant.   1

[Turn over

MARKS | DO NOT WRITE IN THIS MARGIN

7. **(b)** **(i)** **(continued)**

(C) The student burned 1·07 g of methanol and recorded a temperature rise of 23 °C.

Calculate the enthalpy of combustion, in $kJ\,mol^{-1}$, for methanol using the student's results.

3

(ii) The student determined the density of the alcohols by measuring the mass of a volume of each alcohol.

The student's results are shown below.

	Methanol	Ethanol
Volume of alcohol (cm³)	25·0	25·0
Mass of alcohol (g)	19·98	20·05
Density of alcohol (g cm⁻³)		0·802

Calculate the density, in $g\,cm^{-3}$, of methanol.

1

MARKS

DO NOT WRITE IN THIS MARGIN

7. **(continued)**

(c) Methanol is used as a source of hydrogen for fuel cells. The industrial process involves the reaction of methanol with steam.

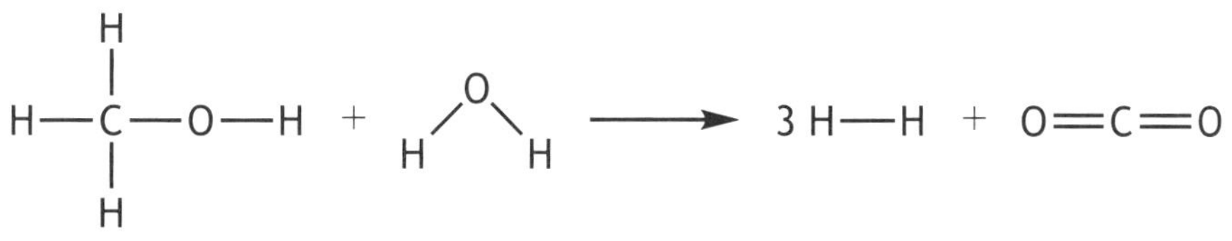

(i) State why it is important for chemists to predict whether reactions in an industrial process are exothermic or endothermic.

1

(ii) Using bond enthalpies from the data booklet, calculate the enthalpy change, in $kJ\,mol^{-1}$, for the reaction of methanol with steam.

2

[Turn over

MARKS

8. Sodium carbonate is used in the manufacture of soaps, glass and paper as well as the treatment of water.

One industrial process used to make sodium carbonate is the Solvay process.

The Solvay process involves several different chemical reactions.

It starts with heating calcium carbonate to produce carbon dioxide, which is transferred to a reactor where it reacts with ammonia and brine. The products of the reactor are solid sodium hydrogencarbonate and ammonium chloride which are passed into a separator.

The sodium hydrogencarbonate is heated to decompose it into the product sodium carbonate along with carbon dioxide and water. To recover ammonia the ammonium chloride from the reactor is reacted with calcium oxide produced by heating the calcium carbonate. Calcium chloride is a by-product of the ammonia recovery process.

(a) Using the information above, complete the flow chart by adding the names of the chemicals involved.　2

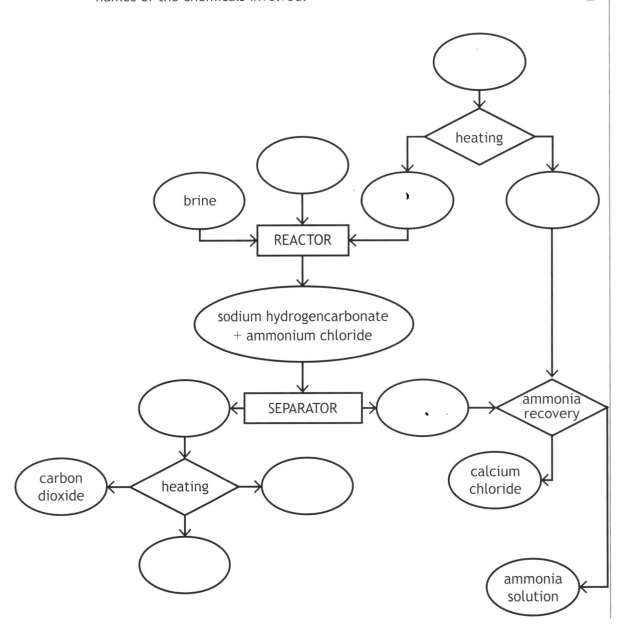

MARKS | DO NOT WRITE IN THIS MARGIN

8.   (continued)

(b) The reaction that produces the solid sodium hydrogencarbonate involves the following equilibrium:

$$HCO_3^-(aq) + Na^+(aq) \rightleftharpoons NaHCO_3(s)$$

Brine is a concentrated sodium chloride solution.

Explain fully why using a concentrated sodium chloride solution encourages production of sodium hydrogencarbonate as a solid.

2

[Turn over

MARKS | DO NOT WRITE IN THIS MARGIN

9. Occasionally, seabirds can become contaminated with hydrocarbons from oil spills. This causes problems for birds because their feathers lose their waterproofing, making the birds susceptible to temperature changes and affecting their buoyancy. If the birds attempt to clean themselves to remove the oil, they may swallow the hydrocarbons causing damage to their internal organs.

Contaminated seabirds can be cleaned by rubbing vegetable oil into their feathers and feet before the birds are rinsed with diluted washing-up liquid.

Using your knowledge of chemistry, comment on the problems created for seabirds by oil spills and the actions taken to treat affected birds.   **3**

[Turn over for Question 10 on *Page twenty-eight*

DO NOT WRITE ON THIS PAGE

MARKS | DO NOT WRITE IN THIS MARGIN

10. Plants require trace metal nutrients, such as zinc, for healthy growth. Zinc ions are absorbed from soil through the plant roots.

The zinc ion concentration in a solution can be found by adding a compound which gives a blue colour to the solution with zinc ions. The concentration of zinc ions is determined by measuring the absorption of light by the blue solution. The higher the concentration of zinc ions in a solution, the more light is absorbed.

A student prepared a stock solution with a zinc ion concentration of $1\,g\,l^{-1}$. Samples from this were diluted to produce solutions of known zinc ion concentration.

(a) The stock solution was prepared by adding $1{\cdot}00\,g$ of zinc metal granules to $20\,cm^3$ of $2\,mol\,l^{-1}$ sulfuric acid in a $1000\,cm^3$ standard flask.

$$Zn(s) \quad + \quad H_2SO_4(aq) \quad \rightarrow \quad ZnSO_4(aq) \quad + \quad H_2(g)$$

The flask was left for 24 hours, without a stopper. The solution was then diluted to $1000\,cm^3$ with water.

(i) **Explain fully** why the flask was left for 24 hours, without a stopper.　　2

(ii) Explain why the student should use deionised water or distilled water, rather than tap water, when preparing the stock solution.　　1

(b) Solutions of known zinc ion concentration were prepared by transferring accurate volumes of the stock solution to standard flasks and diluting with water.

(i) Name the piece of apparatus which should be used to transfer $10\,cm^3$ of stock solution to a standard flask.　　1

MARKS | DO NOT WRITE IN THIS MARGIN

10. (b) (continued)

(ii) Calculate the concentration, in $mg\,l^{-1}$, of the solution prepared by transferring $10\,cm^3$ of the $1\,g\,l^{-1}$ stock solution to a $1000\,cm^3$ standard flask and making up to the mark.

1

(c) The light absorbance of different solutions was measured and the results plotted.

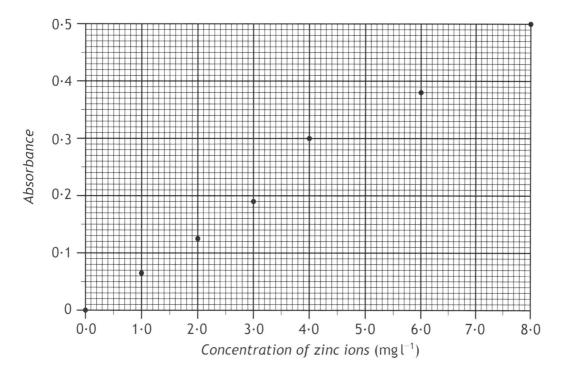

Concentration of zinc ions $(mg\,l^{-1})$

A solution prepared from a soil sample was tested to determine the concentration of zinc ions. The solution had an absorbance of $0\cdot3$.

Determine the concentration, in $mg\,l^{-1}$, of zinc ions in the solution.

1

MARKS | DO NOT WRITE IN THIS MARGIN

11.

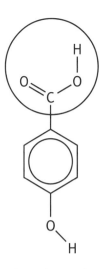

4-hydroxybenzoic acid

4-hydroxybenzoic acid can react with alcohols to form compounds known as parabens.

(a) Name the functional group circled in the structure of 4–hydroxybenzoic acid.

1

(b) Name the type of reaction taking place when parabens are formed.

1

(c) Draw the paraben formed when 4–hydroxybenzoic acid reacts with ethanol.

1

MARKS | DO NOT WRITE IN THIS MARGIN

11.   (continued)

(d) Parabens can be used as preservatives in cosmetics and toiletries.

Parabens are absorbed into the body through the skin. The following table indicates the absorption of some parabens.

Paraben	Absorption ($\mu g\,cm^{-2}$)
Methyl	32·50
Ethyl	20·74
Propyl	11·40
Butyl	7·74
Hexyl	1·60

State a conclusion that can be drawn from the information in the table.   **1**

[Turn over

MARKS | DO NOT WRITE IN THIS MARGIN

12. (a) The concentration of sodium hypochlorite in swimming pool water can be determined by redox titration.

Step 1

A $100 \cdot 0 \, cm^3$ sample from the swimming pool is first reacted with an excess of acidified potassium iodide solution forming iodine.

$$NaOCl(aq) \ + \ 2I^-(aq) \ + \ 2H^+(aq) \ \rightarrow \ I_2(aq) \ + \ NaCl(aq) \ + \ H_2O(\ell)$$

Step 2

The iodine formed in step 1 is titrated using a standard solution of sodium thiosulfate, concentration $0 \cdot 00100 \, mol \, l^{-1}$. A small volume of starch solution is added towards the endpoint.

$$I_2(aq) \ + \ 2Na_2S_2O_3(aq) \ \rightarrow \ 2NaI(aq) \ + \ Na_2S_4O_6(aq)$$

(i) Describe in detail how a burette should be prepared and set up, ready to begin the titration. **3**

(ii) Write the ion-electron equation for the oxidation reaction occurring in step 1. **1**

MARKS | DO NOT WRITE IN THIS MARGIN

12. **(a)** **(continued)**

(iii) Calculate the concentration, in $mol\,l^{-1}$, of sodium hypochlorite in the swimming pool water, if an average volume of $12\cdot4\,cm^3$ of sodium thiosulfate was required.

3

(b) The level of hypochlorite in swimming pools needs to be maintained between 1 and 3 parts per million (1 – 3 ppm).

400 cm^3 of a commercial hypochlorite solution will raise the hypochlorite level of 45 000 litres of water by 1 ppm.

Calculate the volume of hypochlorite solution that will need to be added to an Olympic-sized swimming pool, capacity 2 500 000 litres, to raise the hypochlorite level from 1 ppm to 3 ppm.

2

[Turn over

MARKS | DO NOT WRITE IN THIS MARGIN

12. **(continued)**

(c) The familiar chlorine smell of a swimming pool is not due to chlorine but compounds called chloramines. Chloramines are produced when the hypochlorite ion reacts with compounds such as ammonia, produced by the human body.

$$OCl^-(aq) \; + \; NH_3(aq) \; \rightarrow \; NH_2Cl(aq) \; + \; OH^-(aq)$$
monochloramine

$$OCl^-(aq) \; + \; NH_2Cl(aq) \; \rightarrow \; NHCl_2(aq) \; + \; OH^-(aq)$$
dichloramine

$$OCl^-(aq) \; + \; NHCl_2(aq) \; \rightarrow \; NCl_3(aq) \; + \; OH^-(aq)$$
trichloramine

Chloramines are less soluble in water than ammonia due to the polarities of the molecules, and so readily escape into the atmosphere, causing irritation to the eyes.

(i) Explain the difference in polarities of ammonia and trichloramine molecules.

ammonia   trichloramine

2

MARKS | DO NOT WRITE IN THIS MARGIN

12. (c) (continued)

(ii) Chloramines can be removed from water using ultraviolet light treatment.

One step in the process is the formation of free radicals.

$$NH_2Cl \xrightarrow{UV} \cdot NH_2 \; + \; \cdot Cl$$

State what is meant by the term free radical.   **1**

(iii) Another step in the process is shown below.

$$NH_2Cl \; + \; \cdot Cl \longrightarrow \cdot NHCl \; + \; HCl$$

State the name for this type of step in a free radical reaction.   **1**

[Turn over for Question 13 on *Page thirty-four*

MARKS | DO NOT WRITE IN THIS MARGIN

13. (a) One test for glucose involves Fehling's solution.

Circle the part of the glucose molecule that reacts with Fehling's solution.

1

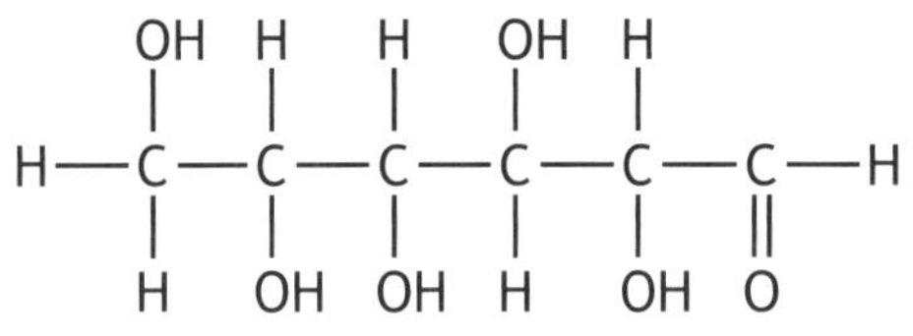

$$\begin{array}{c}
\text{OH} \quad \text{H} \quad\quad \text{H} \quad \text{OH} \quad \text{H} \\
| \quad\quad | \quad\quad | \quad\quad | \quad\quad | \\
\text{H—C—C—C—C—C—C—H} \\
| \quad\quad | \quad\quad | \quad\quad | \quad\quad | \quad\quad \| \\
\text{H} \quad \text{OH} \quad \text{OH} \quad \text{H} \quad \text{OH} \quad \text{O}
\end{array}$$

(b) In solution, sugar molecules exist in an equilibrium in straight-chain and ring forms.

To change from the straight-chain form to the ring form, the oxygen of the hydroxyl on carbon number 5 joins to the carbonyl carbon. This is shown below for glucose.

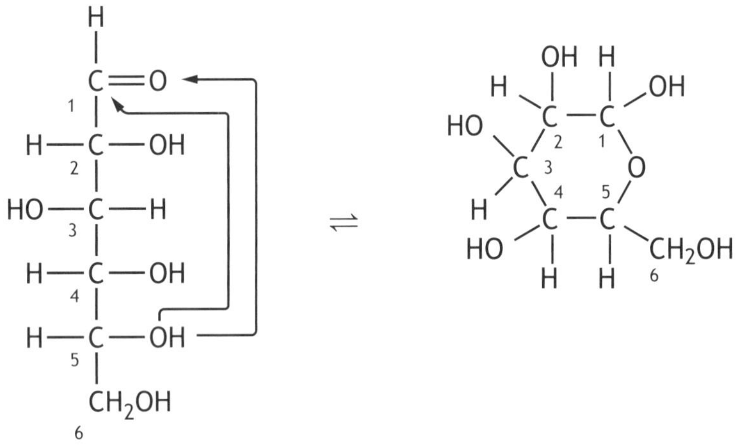

glucose

Draw the structure of a ring form for fructose.

1

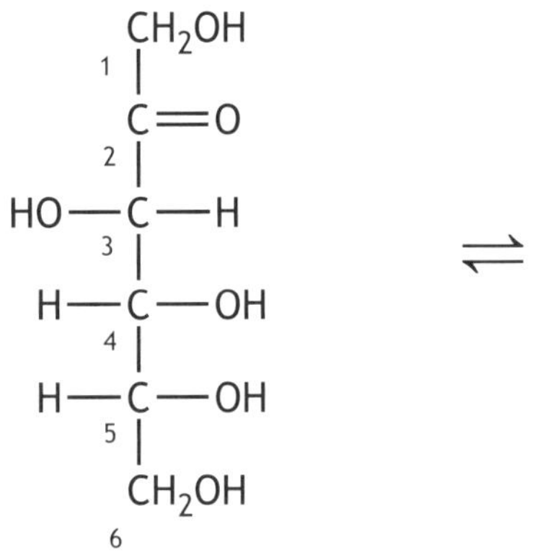

fructose

[END OF QUESTION PAPER]

ADDITIONAL DIAGRAM FOR USE IN QUESTION 3 (a) (i)

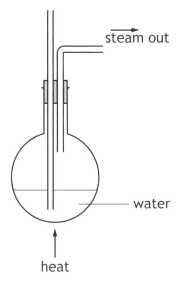

steam out

water

heat

ADDITIONAL SPACE FOR ANSWERS AND ROUGH WORK

ADDITIONAL SPACE FOR ANSWERS AND ROUGH WORK

[BLANK PAGE]

DO NOT WRITE ON THIS PAGE

HIGHER

2016

National Qualifications 2016

X713/76/02

Chemistry
Section 1 — Questions

WEDNESDAY, 18 MAY
9:00 AM – 11:30 AM

Instructions for the completion of Section 1 are given on *Page two* of your question and answer booklet X713/76/01.

Record your answers on the answer grid on *Page three* of your question and answer booklet.

Reference may be made to the Chemistry Higher and Advanced Higher Data Booklet.

Before leaving the examination room you must give your question and answer booklet to the Invigilator; if you do not you may lose all the marks for this paper.

SECTION 1 — 20 marks
Attempt ALL questions

1. Particles with the same electron arrangement are said to be isoelectronic.

    Which of the following compounds contains ions which are isoelectronic?

    A Na_2S

    B $MgCl_2$

    C   KBr

    D $CaCl_2$

2. Which line in the table is correct for the polar covalent bond in hydrogen chloride?

	Relative position of bonding electrons	Dipole notation
A	H —:— Cl	$\delta+$ $\quad$ $\delta-$ H —— Cl
B	H :—— Cl	$\delta+$ $\quad$ $\delta-$ H —— Cl
C	H —:— Cl	$\delta-$ $\quad$ $\delta+$ H —— Cl
D	H :—— Cl	$\delta-$ $\quad$ $\delta+$ H —— Cl

3. Which of the following compounds has the greatest ionic character?

    A   Caesium fluoride

    B   Caesium iodide

    C   Sodium fluoride

    D   Sodium iodide

4. The diagram below shows the energy profiles for a reaction carried out with and without a catalyst.

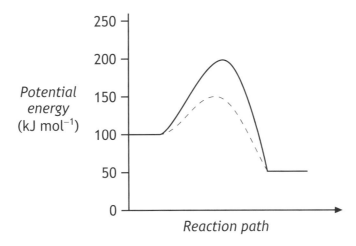

What is the enthalpy change, in kJ mol^{-1}, for the catalysed reaction?

A   −100

B   −50

C   +50

D   +100

5. Limonene is a terpene molecule present in lemons.

limonene

How many isoprene units are joined together in a limonene molecule?

A   1

B   2

C   3

D   4

[Turn over

6. The following molecules give flavour to food.

Which of the following flavour molecules would be most likely to be retained in the food when the food is cooked in water?

A

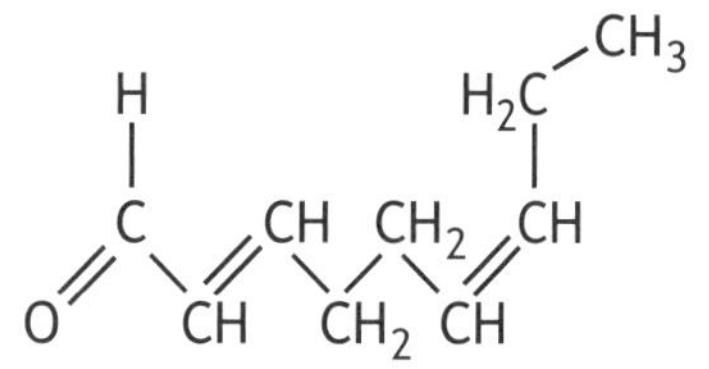

B

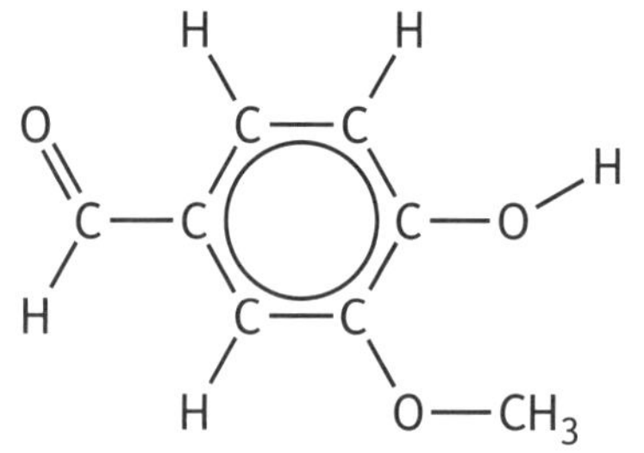

C

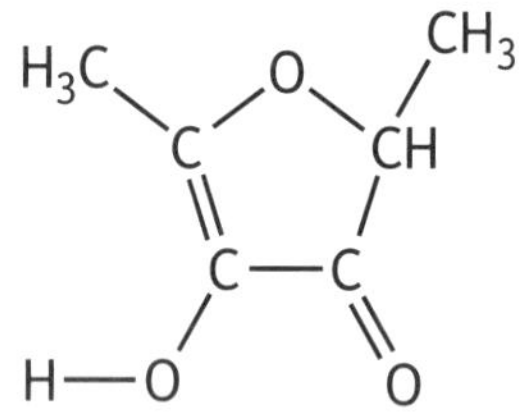

D

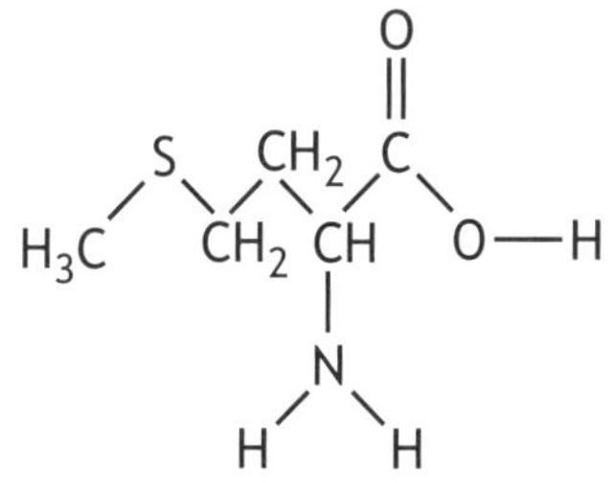

7. vegetable oil $\longrightarrow$ vegetable fat

Which of the following reactions brings about the above change?

A   Hydrolysis

B   Condensation

C   Hydrogenation

D   Dehydrogenation

8. The rate of hydrolysis of protein, using an enzyme, was studied at different temperatures. Which of the following graphs would be obtained?

A

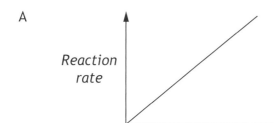

B

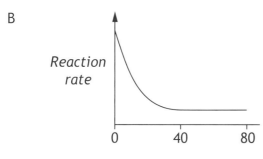

C

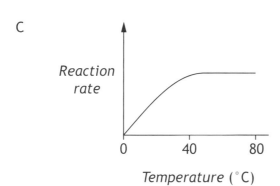

D

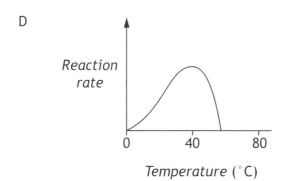

[Turn over

9. Which of the following is the salt of a long-chain fatty acid?

   A   Fat

   B   Oil

   C   Soap

   D   Glycerol

10. Emulsifiers for use in food are commonly made by reacting edible oils with

   A   esters

   B   glycerol

   C   fatty acids

   D   amino acids.

11. The equation for the reduction reaction taking place when ethanal reacts with Tollens' reagent is

   A $Cu^{2+}(aq) + e^- \rightarrow Cu^+(aq)$

   B $Ag^+(aq) + e^- \rightarrow Ag(s)$

   C $Cr_2O_7^{2-}(aq) + 14H^+(aq) + 6e^- \rightarrow 2Cr^{3+}(aq) + 7H_2O(\ell)$

   D $MnO_4^-(aq) + 8H^+(aq) + 5e^- \rightarrow Mn^{2+}(aq) + 4H_2O(\ell)$

12. The name of the compound with structure:

   is

   A   2,3-dimethylpentan-4-one

   B   2,3-dimethylpentan-2-al

   C   3,4-dimethylpentan-2-one

   D   3,4-dimethylpentan-2-al.

13. $CaCO_3(s)$ + $2HNO_3(aq)$ $\rightarrow$ $Ca(NO_3)_2$ (aq) + $CO_2(g)$ + $H_2O(\ell)$

   Mass of 1 mol = 100 g

   Mass of 1 mol = 164 g

   2·00 g of calcium carbonate ($CaCO_3$) was reacted with 200 cm^3 of 0·1 mol l^{-1} nitric acid (HNO_3).

   Take the volume of 1 mole of carbon dioxide to be 24 litres.

   In the reaction

   A $CaCO_3$ is the limiting reactant

   B an excess of 0·1 mol of nitric acid remains at the end of the reaction

   C 1·64 g of calcium nitrate is produced by the reaction

   D 480 cm^3 of carbon dioxide is produced by the reaction.

14. The mean bond enthalpy of a C - F bond is 484 kJ mol^{-1}.

   In which of the processes is ΔH approximately equal to +1936 kJ mol^{-1}?

   A $CF_4(g) \rightarrow C(s) + 2F_2(g)$

   B $CF_4(g) \rightarrow C(g) + 4F(g)$

   C $CF_4(g) \rightarrow C(g) + 2F_2(g)$

   D $CF_4(g) \rightarrow C(s) + 4F(g)$

15. In a reversible reaction, equilibrium is reached when

   A molecules of reactants cease to change into molecules of products

   B the concentrations of reactants and products are equal

   C the concentrations of reactants and products are constant

   D the activation energy of the forward reaction is equal to that of the reverse reaction.

16. Which of the following equations represents the enthalpy of combustion of propane?

   A $C_3H_8(g) + 5O_2(g) \rightarrow 3CO_2(g) + 4H_2O(\ell)$

   B $C_3H_8(g) + \frac{7}{2}O_2(g) \rightarrow 3CO(g) + 4H_2O(\ell)$

   C $C_3H_8(g) + 3O_2(g) \rightarrow 3CO_2(g) + 4H_2(g)$

   D $C_3H_8(g) + \frac{3}{2}O_2(g) \rightarrow 3CO(g) + 4H_2(g)$

[Turn over

17. An oxidising agent

    A   gains electrons and is oxidised

    B   loses electrons and is oxidised

    C   gains electrons and is reduced

    D   loses electrons and is reduced.

18. During a redox process in acid solution, chlorate ions, ClO_3^-(aq), are converted into chlorine, Cl_2(g).

$$ClO_3^-(aq) \rightarrow Cl_2(g)$$

    The numbers of H^+(aq) and $H_2O(\ell)$ required to balance the ion-electron equation for the formation of 1 mol of Cl_2(g) are, respectively

    A   3 and 6

    B   6 and 3

    C   6 and 12

    D   12 and 6.

19. Which of the following ions could be used to oxidise iodide ions to iodine?

$$2I^-(aq) \rightarrow I_2(s) + 2e^-$$

    A SO_4^{2-}(aq)

    B SO_3^{2-}(aq)

    C Cr^{3+}(aq)

    D $Cr_2O_7^{2-}$(aq)

20.

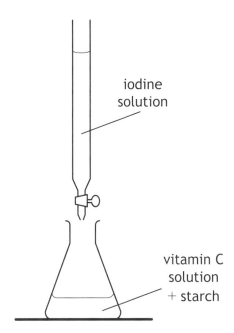

A student was carrying out a titration to establish the concentration of vitamin C using iodine solution.

Which of the following would help the student achieve a precise end-point?

A   Placing a white tile underneath the conical flask

B   Using the bottom of the meniscus when reading the burette

C   Repeating titrations

D   Carrying out a rough titration first

[END OF SECTION 1. NOW ATTEMPT THE QUESTIONS IN SECTION 2
OF YOUR QUESTION AND ANSWER BOOKLET.]

[BLANK PAGE]

DO NOT WRITE ON THIS PAGE

FOR OFFICIAL USE

National
Qualifications
2016

Mark

X713/76/01

Chemistry
Section 1 — Answer Grid and Section 2

WEDNESDAY, 18 MAY

9:00 AM — 11:30 AM

Fill in these boxes and read what is printed below.

Full name of centre

Town

Forename(s)

Surname

Number of seat

Date of birth

Day   Month   Year

Scottish candidate number

Total marks — 100

SECTION 1 — 20 marks

Attempt ALL questions.

Instructions for completion of Section 1 are given on *Page two*.

SECTION 2 — 80 marks

Attempt ALL questions

Reference may be made to the Chemistry Higher and Advanced Higher Data Booklet.

Write your answers clearly in the spaces provided in this booklet. Additional space for answers and rough work is provided at the end of this booklet. If you use this space you must clearly identify the question number you are attempting. Any rough work must be written in this booklet. You should score through your rough work when you have written your final copy.

Use **blue** or **black** ink.

Before leaving the examination room you must give this booklet to the Invigilator; if you do not, you may lose all the marks for this paper.

SECTION 1 — 20 marks

The questions for Section 1 are contained in the question paper X713/76/02.
Read these and record your answers on the answer grid on *Page three* opposite.
Use **blue** or **black** ink. Do NOT use gel pens or pencil.

1. The answer to each question is **either** A, B, C or D. Decide what your answer is, then fill in the appropriate bubble (see sample question below).

2. There is **only one correct** answer to each question.

3. Any rough working should be done on the additional space for answers and rough work at the end of this booklet.

Sample Question

To show that the ink in a ball-pen consists of a mixture of dyes, the method of separation would be:

    A   fractional distillation

    B   chromatography

    C   fractional crystallisation

    D   filtration.

The correct answer is **B**—chromatography. The answer **B** bubble has been clearly filled in (see below).

Changing an answer

If you decide to change your answer, cancel your first answer by putting a cross through it (see below) and fill in the answer you want. The answer below has been changed to **D**.

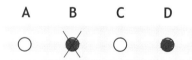

If you then decide to change back to an answer you have already scored out, put a tick (✓) to the **right** of the answer you want, as shown below:

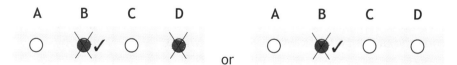

SECTION 1 — Answer Grid

	A	B	C	D
1	○	○	○	○
2	○	○	○	○
3	○	○	○	○
4	○	○	○	○
5	○	○	○	○
6	○	○	○	○
7	○	○	○	○
8	○	○	○	○
9	○	○	○	○
10	○	○	○	○
11	○	○	○	○
12	○	○	○	○
13	○	○	○	○
14	○	○	○	○
15	○	○	○	○
16	○	○	○	○
17	○	○	○	○
18	○	○	○	○
19	○	○	○	○
20	○	○	○	○

[Turn over

[BLANK PAGE]

DO NOT WRITE ON THIS PAGE

[Turn over for next question

DO NOT WRITE ON THIS PAGE

SECTION 2 — 80 marks

Attempt ALL questions

1. Hydrogen peroxide gradually decomposes into water and oxygen, according to the following equation.

$$2H_2O_2(aq) \rightarrow 2H_2O(\ell) + O_2(g)$$

(a) At room temperature, the reaction is very slow. It can be speeded up by heating the reaction mixture.

State why increasing the temperature causes an increase in reaction rate.

1

(b) (i) The reaction can also be speeded up by adding a catalyst, such as manganese dioxide.

To determine the rate of the reaction, the volume of gas produced in a given time can be measured.

Complete the diagram below to show how the gas produced can be collected and measured.

1

(An additional diagram, if required, can be found on *Page thirty-eight*.)

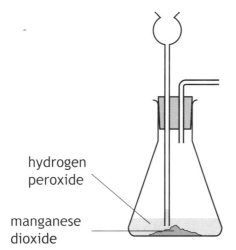

hydrogen
peroxide

manganese
dioxide

MARKS | DO NOT WRITE IN THIS MARGIN

1. (b) (continued)

(ii) The concentration of hydrogen peroxide is often described as a volume strength. This relates to the volume of oxygen that can be produced from a hydrogen peroxide solution.

volume of oxygen produced	=	volume strength	×	volume of hydrogen peroxide solution

In an experiment, $74\,cm^3$ of oxygen was produced from $20\,cm^3$ of hydrogen peroxide solution.

Calculate the volume strength of the hydrogen peroxide.   **1**

(c) Hydrogen peroxide can react with potassium iodide to produce water and iodine.

A student carried out an experiment to investigate the effect of changing the concentration of potassium iodide on reaction rate. The results are shown below.

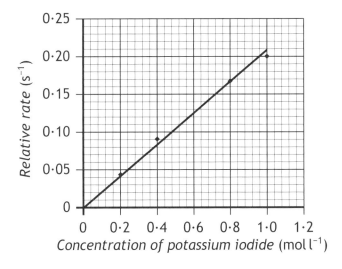

Calculate the time taken, in s, for the reaction when the concentration of potassium iodide used was $0.6\,mol\,l^{-1}$.   **1**

[Turn over

MARKS | DO NOT WRITE IN THIS MARGIN

2.   (a)   Graph 1 shows the sizes of atoms and ions for elements in the third period of the Periodic Table.

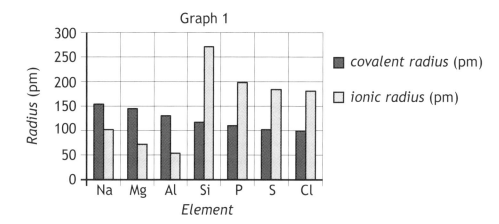

The covalent radius is a measure of the size of an atom.

   (i)   Explain why covalent radius decreases across the period from sodium to chlorine.

1

   (ii)   Explain **fully** why the covalent radius of sodium is larger than the ionic radius of sodium.

2

MARKS | DO NOT WRITE IN THIS MARGIN

2. (continued)

(b) Graph 2 shows the first and second ionisation energies of elements in Group 1 of the Periodic Table.

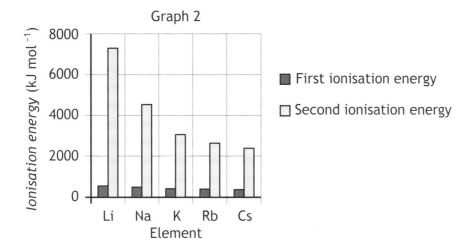

Graph 2

(i) Explain why the first ionisation energy decreases going down Group 1.

1

(ii) Explain **fully** why the second ionisation energy is much greater than the first ionisation energy for Group 1 elements.

2

[Turn over

MARKS | DO NOT WRITE IN THIS MARGIN

2. **(continued)**

(c) The lattice enthalpy is the energy needed to completely separate the ions in one mole of an ionic solid.

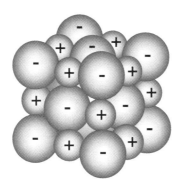

Table 1 shows the size of selected ions.

Table 1

Ion	Li^+	Na^+	K^+	Rb^+	F^-	Cl^-
Ionic radius (pm)	76	102	138	152	133	181

Table 2 shows the lattice enthalpies, in $kJ\,mol^{-1}$, for some Group 1 halides.

Table 2

Ions	F^-	Cl^-
Li^+	1030	834
Na^+	910	769
K^+	808	701
Rb^+		658

(i) Predict the lattice enthalpy, in $kJ\,mol^{-1}$, for rubidium fluoride.   1

(ii) Write a general statement linking lattice enthalpy to ionic radii.   1

3. Phosphine (PH_3) is used as an insecticide in the storage of grain.

Phosphine can be produced by the reaction of water with aluminium phosphide

$$AlP(s) + 3H_2O(\ell) \longrightarrow PH_3(g) + Al(OH)_3(aq)$$

(a) State the type of bonding and structure in phosphine.   **1**

(b) 2·9 kg of aluminium phosphide were used in a phosphine generator.

Calculate the volume of phosphine gas, in litres, that would have been produced.

(Take the volume of 1 mole of phosphine to be 24 litres).   **2**

(c) Carbon dioxide is fed into the phosphine generator to keep the phosphine concentration less than 2·6%. Above this level phosphine can ignite due to the presence of diphosphane, $P_2H_4(g)$, as an impurity.

Draw a structural formula for diphosphane.   **1**

[Turn over

4. The viscosity of alcohols depends on a number of factors:

- the strength of intermolecular forces
- the size of the molecule
- temperature

These factors can be investigated using alcohols and apparatus from the lists below.

Alcohols	Apparatus
methanol	beakers
ethanol	funnels
propan-1-ol	burettes
ethane-1,2-diol	measuring cylinders
butan-1-ol	plastic syringes
propane-1,3-diol	glass tubing
pentan-1-ol	stoppers
propane-1,2,3-triol	timer
	metre stick
	ball bearing
	clamp stands
	kettle
	thermometer

Using your knowledge of chemistry, identify the alcohols and apparatus that you would select and describe how these could be used to investigate one, or more, of the factors affecting the viscosity of alcohols.

3

4.   (continued)   Answer space

[Turn over

MARKS | DO NOT WRITE IN THIS MARGIN

5. When fats and oils are hydrolysed, mixtures of fatty acids are obtained.

    (a) Name the other product obtained in this reaction.   **1**

    (b) The table below shows the percentage composition of the fatty acid mixtures obtained by hydrolysis of coconut oil and olive oil.

Class of fatty acids produced on hydrolysis	Name of oil	
	Coconut oil	Olive oil
Saturated	91	14
Monounsaturated	6	72
Polyunsaturated	3	14

        (i) One of the fatty acids produced by the hydrolysis of olive oil is linoleic acid, $C_{17}H_{31}COOH$.

             State the class of fatty acid to which linoleic acid belongs.   **1**

        (ii) Hydrolysed coconut oil contains the fatty acid, caprylic acid, with the formula $CH_3(CH_2)_6COOH$.

             State the systematic name for caprylic acid.   **1**

    (c) The degree of unsaturation of oil can be tested by adding drops of bromine solution to the oil. Bromine adds across carbon to carbon double bonds in the fatty acid chains.

MARKS

DO NOT WRITE IN THIS MARGIN

5. **(c)** **(continued)**

The following apparatus can be used to compare the degree of unsaturation of different oils.

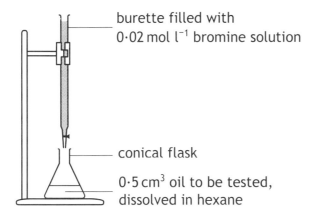

burette filled with
0·02 mol l^{-1} bromine solution

conical flask

0·5 cm^3 oil to be tested,
dissolved in hexane

    (i) Describe how this apparatus could be used to show that olive oil has a greater degree of unsaturation than coconut oil.

2

    (ii) Suggest why hexane is used as the solvent, rather than water.

1

    (iii) Coconut oil has a melting point of 25 °C. Olive oil has a melting point of −6 °C.

Give two reasons why coconut oil has a higher melting point than olive oil.

2

[Turn over

MARKS | DO NOT WRITE IN THIS MARGIN

6. Peptide molecules can be classified according to the number of amino acid units joined by peptide bonds in the molecule.

Type of peptide	Example of amino acid sequence
dipeptide	aspartic acid-phenylalanine
tripeptide	isoleucine-proline-proline
tetrapeptide	lysine-proline-proline-arginine
pentapeptide	serine-glycine-tyrosine-alanine-leucine
	alanine-glycine-valine-proline-tyrosine-serine
polypeptide	many amino acids

(a) Complete the table to identify the type of peptide with the following amino acid sequence

   alanine-glycine-valine-proline-tyrosine-serine   1

(b) Partial hydrolysis of another pentapeptide molecule gave a mixture of three smaller peptide molecules with the following amino acid sequences.

   leucine-glycine-valine

   isoleucine-leucine

   glycine-valine-serine

   Write the amino acid sequence for the original pentapeptide molecule.   1

(c) Some amino acids needed to form polypeptides cannot be produced in the human body.

   State the term used to describe amino acids that the body cannot make.   1

MARKS | DO NOT WRITE IN THIS MARGIN

6.   (continued)

(d) Paper chromatography is often used to analyse the mixtures of amino acids produced when peptides are broken down.

On a chromatogram, the retention factor R_f, for a substance can be a useful method of identifying the substance.

$$R_f = \frac{\text{distance moved by the substance}}{\text{maximum distance moved by the solvent}}$$

The structure of the pentapeptide methionine enkephalin was investigated.

A sample of the pentapeptide was completely hydrolysed into its constituent amino acids and this amino acid mixture was applied to a piece of chromatography paper and placed in a solvent.

The chromatogram obtained is shown below.

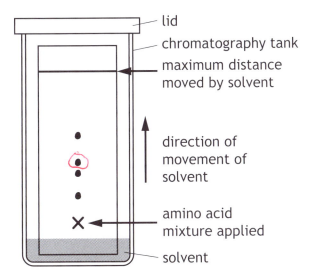

(i) Suggest why only four spots were obtained on the chromatogram of the hydrolysed pentapeptide.   1

(ii) It is known that this amino acid mixture contains the amino acid methionine. The R_f value for methionine in this solvent is 0·40.

Draw a circle around the spot on the chromatogram that corresponds to methionine.   1

[Turn over

MARKS | DO NOT WRITE IN THIS MARGIN

6. (continued)

(e) Over the last decade several families of extremely stable peptide molecules have been discovered, where the peptide chain forms a ring.

(i) A simple cyclic dipeptide is shown.

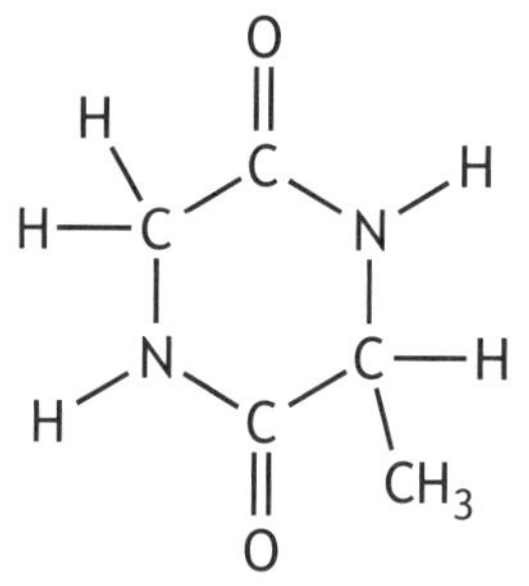

Draw a structural formula for one of the amino acids that would be formed on complete hydrolysis of the above cyclic dipeptide.

1

(ii) Alpha-amanitin is a highly toxic cyclic peptide found in death cap mushrooms. The lethal dose for humans is 100 mg per kg of body mass.

1·0 g of death cap mushrooms contains 250 mg of alpha-amanitin.

Calculate the minimum mass of death cap mushrooms that would contain the lethal dose for a 75 kg adult.

2

[Turn over for next question

DO NOT WRITE ON THIS PAGE

MARKS | DO NOT WRITE IN THIS MARGIN

7. Modern shellac nail varnishes are more durable and so last longer than traditional nail polish.

The shellac nail varnish is applied in thin layers to the nails and then the fingers are placed under a UV lamp.

(a) The Skin Care Foundation has recommended that a sun-block is applied to the fingers and hand before using the lamp.

Suggest why the Skin Care Foundation makes this recommendation.   1

(b) A *free radical* chain reaction takes place and the varnish hardens.

    (i) State what is meant by the term *free radical*.   1

MARKS | DO NOT WRITE IN THIS MARGIN

7. **(b) (continued)**

(ii) The shellac nail varnish contains a mixture of ingredients that take part in the free radical chain reaction.

One of the steps in the free radical chain reaction is:

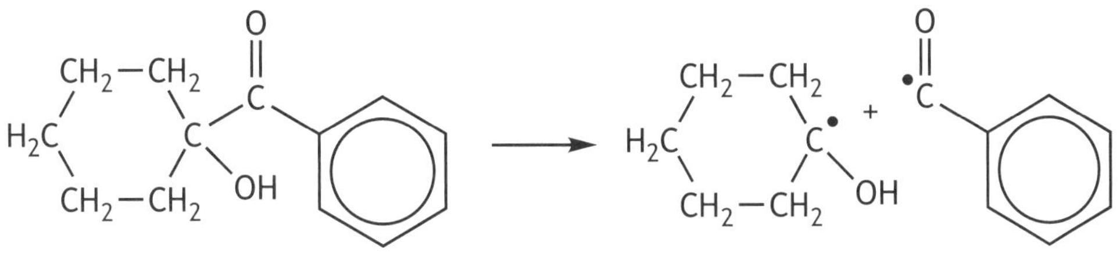

State the term used to describe this type of step in a free radical chain reaction. **1**

(iii) During the free radical chain reaction small molecules join to form large chain molecules.

One example of a small molecule used is

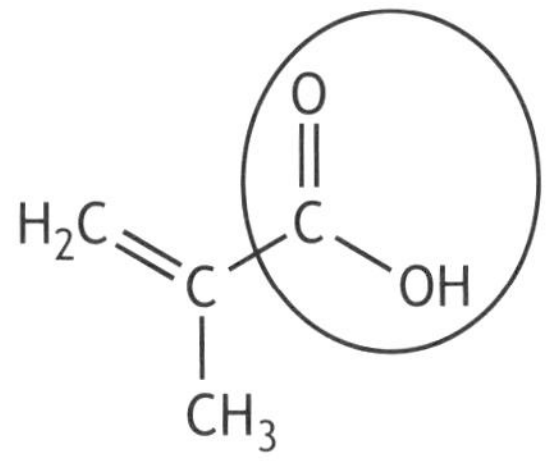

Name the functional group circled above. **1**

(iv) Alcohol wipes are used to finish the varnishing treatment. Alcohol wipes contain the alcohol propan-2-ol.

State why propan-2-ol can be described as a secondary alcohol. **1**

[Turn over

MARKS | DO NOT WRITE IN THIS MARGIN

7. **(continued)**

(c) Traditional nail varnishes use ethyl ethanoate and butyl ethanoate as solvents.

(i) Draw a structural formula for butyl ethanoate.   **1**

(ii) Ethyl ethanoate can be made in the laboratory using the following apparatus.

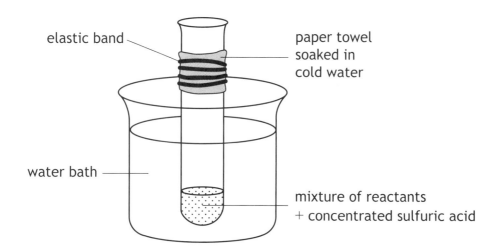

elastic band

paper towel soaked in cold water

water bath

mixture of reactants + concentrated sulfuric acid

Suggest why a wet paper towel is wrapped around the test tube.   **1**

7. (c) (continued)

(iii) A student used 2·5 g of ethanol and a slight excess of ethanoic acid to produce 2·9 g of ethyl ethanoate.

ethanol + ethanoic acid ⇌ ethyl ethanoate + water

mass of
one mole
= 46·0 g

mass of
one mole
= 88·0 g

(One mole of ethanol reacts with one mole of ethanoic acid to produce one mole of ethyl ethanoate.)

Calculate the percentage yield of ethyl ethanoate.   2

(iv) Name the type of reaction that takes place during the formation of ethyl ethanoate.   1

[Turn over

MARKS

8. Methanol (CH_3OH) is an important chemical in industry.

    (a) Methanol is produced from methane in a two-step process.

    In step 1, methane is reacted with steam as shown.

    Step 1: $CH_4(g) + H_2O(g) \rightleftharpoons 3H_2(g) + CO(g)$ $\Delta H = +210\,kJ\,mol^{-1}$

    In step 2, hydrogen reacts with carbon monoxide.

    Step 2: $2H_2(g) + CO(g) \rightleftharpoons CH_3OH(g)$ $\Delta H = -91\,kJ\,mol^{-1}$

    Complete the table to show the most favourable conditions to maximise
    the yield for each step.   **2**

	Temperature (High/Low)	Pressure (High/Low)
Step 1		
Step 2		

    (b) Methanol reacts with compound X, in an addition reaction, to form
    methyl tertiary-butyl ether, an additive for petrol.

    $CH_3OH(g) + X \longrightarrow$

    methyl tertiary-butyl ether

    (i) Suggest a structure for compound X.   **1**

    (ii) The atom economy of this reaction is 100%.

    Explain what this means.   **1**

MARKS | DO NOT WRITE IN THIS MARGIN

8. (continued)

(c) Methanol can be converted to methanal as shown.

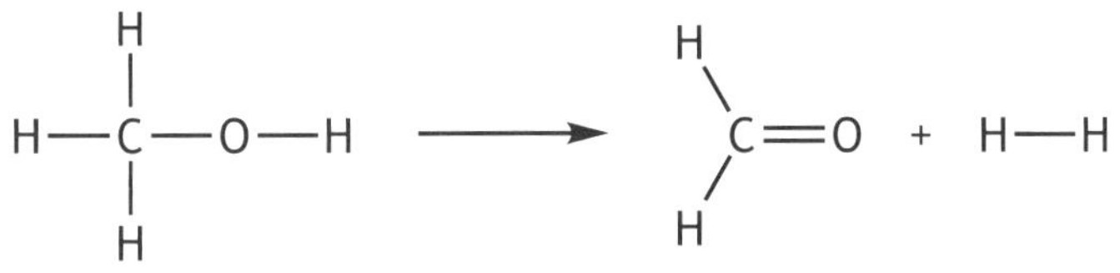

Using bond enthalpy and mean bond enthalpy values from the data booklet, calculate the enthalpy change, in $kJ\,mol^{-1}$, for the reaction.   2

[Turn over

9. A group of students carried out an investigation into the energy changes that take place when metal hydroxides dissolve in water.

The following apparatus was used as a simple calorimeter to determine the change in temperature.

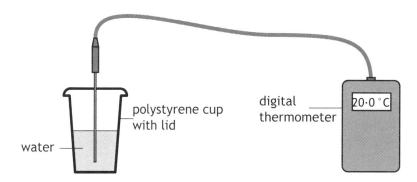

The experiment was carried out as follows.

Step 1: 100 cm³ of deionised water was added to the cup.

Step 2: The stop-clock was started, the water stirred continuously and the temperature recorded every 20 seconds.

Step 3: After 60 seconds, an accurately weighed mass of the metal hydroxide was added to the water and the temperature recorded every 20 seconds.

Graph 1 shows the group's results for lithium hydroxide.

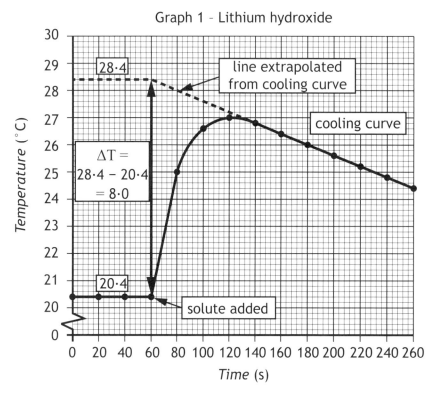

Graph 1 – Lithium hydroxide

The heat energy transferred to the water can be calculated as shown.

$$E_h = cm\Delta T$$
$$= 4.18 \times 0.10 \times 8.0$$
$$= 3.3 \, kJ$$

MARKS | DO NOT WRITE IN THIS MARGIN

9. (continued)

(a) The experiment was repeated using sodium hydroxide.

Graph 2 shows the results of this experiment.

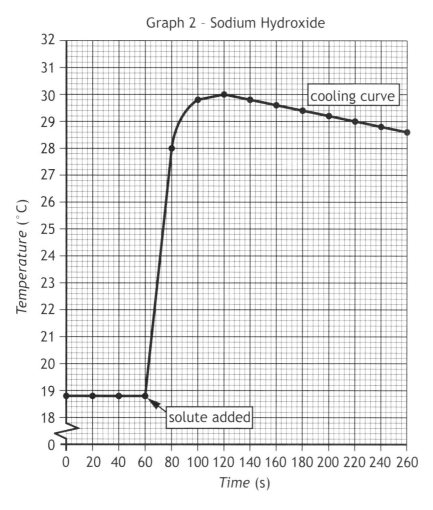

Graph 2 – Sodium Hydroxide

(i) Using Graph 2 calculate the heat energy transferred to the water, in kJ, when the sodium hydroxide dissolved.

2

(ii) Suggest why the experiment was carried out in a polystyrene cup with a lid.

1

[Turn over

MARKS | DO NOT WRITE IN THIS MARGIN

9.   (a) (continued)

   (iii)   **In another experiment** the students found that $5.61\,g$ of potassium hydroxide (KOH) released $5.25\,kJ$ of heat energy on dissolving.

   Use this information to calculate the energy released, in $kJ\,mol^{-1}$, when one mole of potassium hydroxide dissolves in water.   **1**

   (b)   Calcium hydroxide solution can be formed by adding calcium metal to excess water.

   Solid calcium hydroxide would form if the exact molar ratio of calcium to water is used. The equation for the reaction is

   $$Ca(s) + 2H_2O(\ell) \rightarrow Ca(OH)_2(s) + H_2(g)$$

   Calculate the enthalpy change, in $kJ\,mol^{-1}$, for the reaction above by using the data shown below.

   $$H_2(g) + \tfrac{1}{2}O_2(g) \rightarrow H_2O(\ell) \qquad \Delta H = -286\,kJ\,mol^{-1}$$

   $$Ca(s) + O_2(g) + H_2(g) \rightarrow Ca(OH)_2(s) \qquad \Delta H = -986\,kJ\,mol^{-1}$$ **2**

MARKS | DO NOT WRITE IN THIS MARGIN

10. The chemical industry creates an immense variety of products which impact on virtually every aspect of our lives. Industrial scientists, including chemical engineers, production chemists and environmental chemists, carry out different roles to maximise the efficiency of industrial processes.

Using your knowledge of chemistry, comment on what industrial scientists can do to maximise profit from industrial processes and minimise impact on the environment.

3

[Turn over

MARKS | DO NOT WRITE IN THIS MARGIN

11. Soft drinks contain a variety of sugars. A student investigated the sugar content of a soft drink.

    (a) The density of the soft drink can be used to estimate its total sugar concentration. Solutions of different sugars, with the same concentration, have similar densities.

    The first experiment was to determine the total sugar concentration of the soft drink by comparing the density of the drink with the density of standard sucrose solutions.

        (i) This firstly involved producing standard sucrose solutions of different concentrations.

        The standard sucrose solutions were made up in volumetric flasks.

        Draw a diagram of a volumetric flask.   **1**

        (ii) The density of each standard sucrose solution was then determined. In order to determine the density of each solution, the student accurately measured the mass of $10 \cdot 0 \, cm^3$ of each sucrose solution.

        Describe **fully** a method that the student could have used to accurately measure the mass of $10 \cdot 0 \, cm^3$ of each sucrose solution.   **2**

MARKS | DO NOT WRITE IN THIS MARGIN

11. **(a)** **(continued)**

(iii) The results that the student obtained for the density of the standard solutions of sucrose are shown in the table.

% Concentration of sucrose solution	Density of sucrose solution ($g\,cm^{-3}$)
0·0	1·00
5·0	1·10
10·0	1·19
15·0	1·31
20·0	1·41

Draw a line graph using the student's results.   2

(Additional graph paper, if required, can be found on *Page thirty-eight*.)

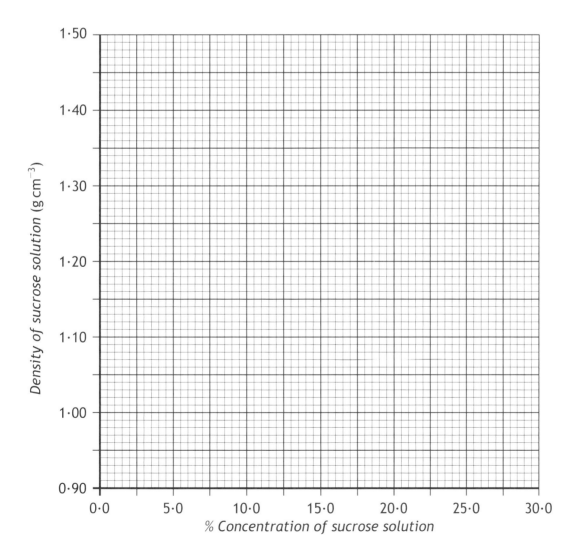

[Turn over

MARKS | DO NOT WRITE IN THIS MARGIN

11. (a) (continued)

(iv) The student used the line graph to obtain the relationship between the concentration of sugars in solution and the density of the solution.

This equation shows the relationship.

density of sugar in $g\,cm^{-3}$ = $(0.0204 \times \%$ concentration of sugars in solution$) + 1.00$

The student then determined the density of a soft drink. In order to ensure that the drink was flat, all the gas had been allowed to escape.

(A) Suggest a reason why the soft drink needed to be flat before its density was determined.   1

(B) The soft drink tested had a density of $1.07\,g\,cm^{-3}$.

Using the equation, calculate the % concentration of sugars present in the soft drink.   1

(v) A different soft drink is found to contain 10·6 grams of sugar in $100\,cm^3$.

Calculate the total mass of sugar present, in grams, in a $330\,cm^3$ can of this soft drink.   1

(b) The second experiment in the investigation was to determine the concentration of specific types of sugar called reducing sugars. This was carried out by titration with Fehling's solution.

(i) Reducing sugars contain an aldehyde functional group.

Draw this functional group.   1

MARKS | DO NOT WRITE IN THIS MARGIN

11. (b) (continued)

(ii) The overall reaction that occurs with Fehling's solution and a reducing sugar is shown.

$$C_6H_{12}O_6 + 2Cu^{2+} + H_2O \rightarrow C_6H_{12}O_7 + 2Cu^+ + 2H^+$$

reducing   Fehling's
sugar   solution

Write the ion-electron equation for the oxidation reaction.   1

(iii) State the colour change that would be observed when reducing sugars are reacted with Fehling's solution.   1

(iv) For the titrations, the student diluted the soft drink to improve the accuracy of results.

$25.0\,cm^3$ samples of the diluted soft drink were titrated with Fehling's solution which had a Cu^{2+} concentration of $0.0250\,mol\,l^{-1}$.

The average volume of Fehling's solution used in the titrations was $19.8\,cm^3$.

$$C_6H_{12}O_6 + 2Cu^{2+} + H_2O \rightarrow C_6H_{12}O_7 + 2Cu^+ + 2H^+$$

reducing   Fehling's
sugar   solution

Calculate the concentration, in $mol\,l^{-1}$, of reducing sugars present in the diluted sample of the soft drink.   3

[Turn over

MARKS

12. (a) The table shows the boiling points and structures of some isomers with molecular formula $C_6H_{12}O_2$.

Isomer	Structure	Boiling point (°C)
1		205
2		201
3		187
4		132
5		125
6		119
7		126
8		98

MARKS

12. (a) (continued)

(i)   Name the intermolecular force which accounts for the higher boiling points of isomers 1, 2 and 3.   **1**

(ii) Using the information in the table, describe **two** ways in which differences in structure affect the boiling points of isomeric esters 4–8.   **2**

(iii) Predict the boiling point, in °C, for the isomer shown below.   **1**

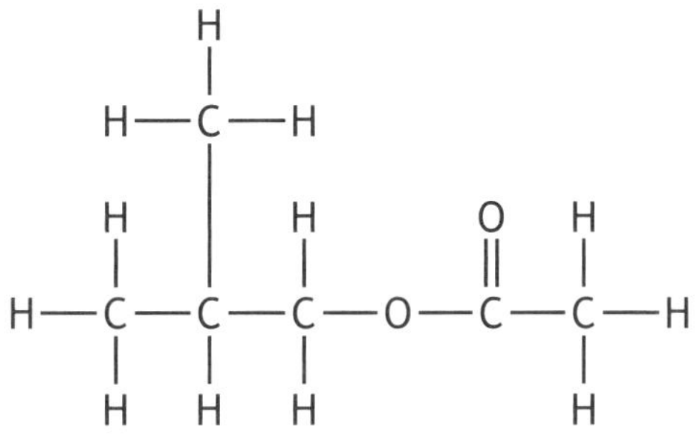

```
              H
              |
          H — C — H
              |
      H       H           O   H
      |       |           ||  |
  H — C — C — C — O — C — C — H
      |   |   |           |
      H   H   H           H
```

[Turn over

12.   (continued)

(b) Carbon-13 NMR spectroscopy is a technique that can be used in chemistry to determine the structure of organic molecules such as esters.

In a carbon-13 NMR spectrum, a carbon atom in a molecule is identified by its **chemical shift**. This value depends on the other atoms bonded to the carbon atom, which is known as the "chemical environment" of the carbon-13 atom.

Carbon-13 chemical shift values are shown in the table below.

The carbon-13 atom in each chemical environment has been circled.

Chemical environment	Chemical shift (ppm)
	25–35
	16–25
	50–90
	10–15
	20–50
	170–185

The **number** of peaks in a carbon-13 NMR spectrum corresponds to the number of carbon atoms in different chemical environments within the molecule.

The **position** of a peak (the chemical shift) indicates the type of carbon atom.

MARKS | DO NOT WRITE IN THIS MARGIN

12.   (b)   (continued)

The spectrum for ethyl ethanoate is shown below.

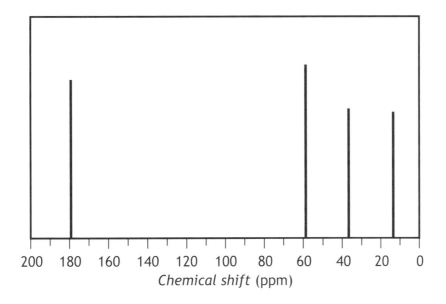

(i)   Label each peak in the ethyl ethanoate spectrum with a number to match the carbon atom in ethyl ethanoate, shown below.   1

(ii)   Determine the number of peaks that would be seen in the carbon-13 NMR spectrum for the ester shown below.   1

Number of peaks in carbon-13 NMR spectrum

= _____

[END OF QUESTION PAPER]

MARKS | DO NOT WRITE IN THIS MARGIN

ADDITIONAL DIAGRAM FOR USE IN QUESTION 1 (b) (i)

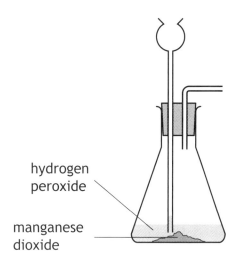

hydrogen peroxide

manganese dioxide

ADDITIONAL GRAPH PAPER FOR QUESTION 11 (a) (iii)

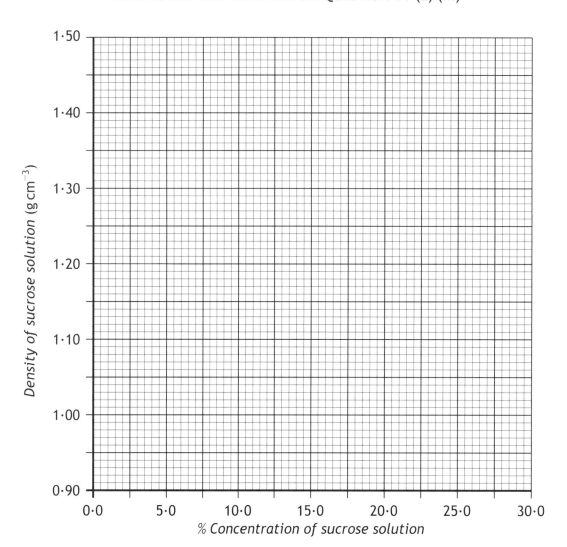

Density of sucrose solution (g cm^{-3})

% Concentration of sucrose solution

ADDITIONAL SPACE FOR ANSWERS AND ROUGH WORK

ADDITIONAL SPACE FOR ANSWERS AND ROUGH WORK

National
Qualifications
2017

X713/76/02

Chemistry
Section 1 — Questions

MONDAY, 8 MAY

9:00 AM – 11:30 AM

Instructions for the completion of Section 1 are given on *Page two* of your question and answer booklet X713/76/01.

Record your answers on the answer grid on *Page three* of your question and answer booklet.

You may refer to the Chemistry Data Booklet for Higher and Advanced Higher.

Before leaving the examination room you must give your question and answer booklet to the Invigilator; if you do not, you may lose all the marks for this paper.

SECTION 1 — 20 marks

Attempt ALL questions

1. Which of the following bonds is the **least** polar?

   A C—I

   B C—F

   C C—Cl

   D C—Br

2. Which of the following compounds would be the **most water** soluble?

   A 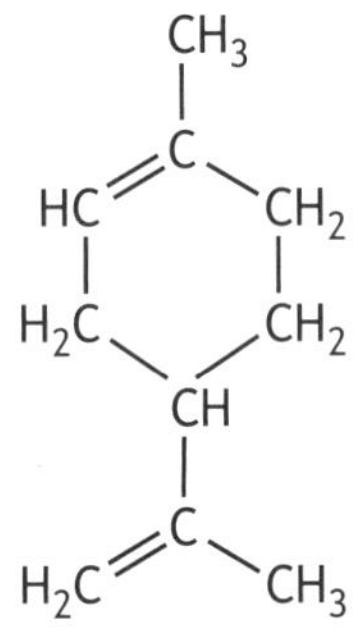

   B 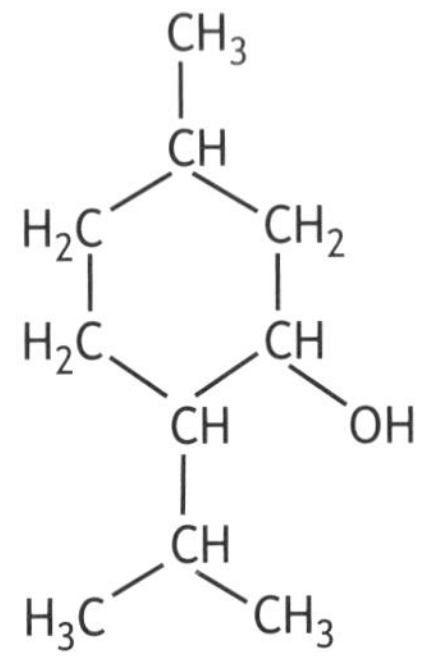

   C 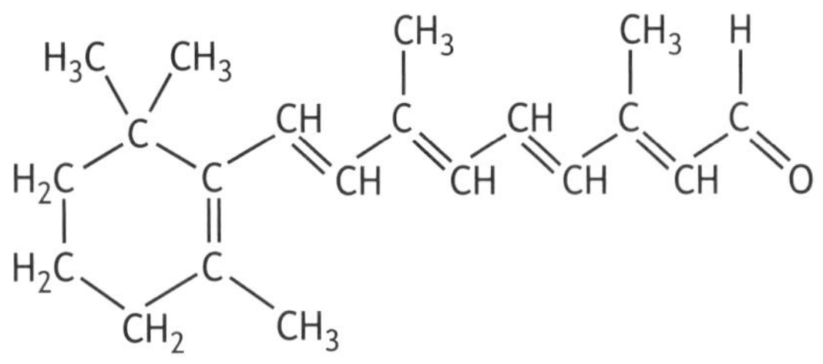

   D 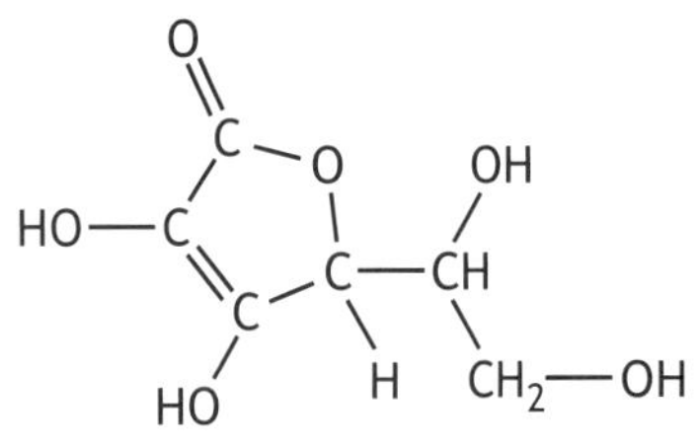

3. Which of the following atoms has the greatest attraction for bonding electrons?

    A   Sulfur

    B   Silicon

    C   Nitrogen

    D   Hydrogen

4. Which type of structure is found in phosphorus?

    A   Covalent network

    B   Covalent molecular

    C   Monatomic

    D   Metallic lattice

5. The polarity of molecules can be investigated using a charged rod. The charged rod will attract a stream of polar liquid flowing from a burette.

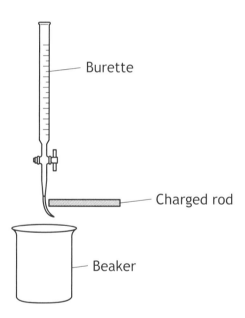

    Which of the following liquids would **not** be attracted?

    A   Water

    B   Propanone

    C   Propanol

    D   Hexane

[Turn over

6. $xP_2H_4 + yO_2 \rightarrow P_4O_{10} + zH_2O$

The equation is balanced when

A $x = 1, y = 5, z = 4$

B $x = 4, y = 6, z = 2$

C $x = 2, y = 7, z = 4$

D $x = 2, y = 5, z = 2$

7. What is the systematic name for the compound below?

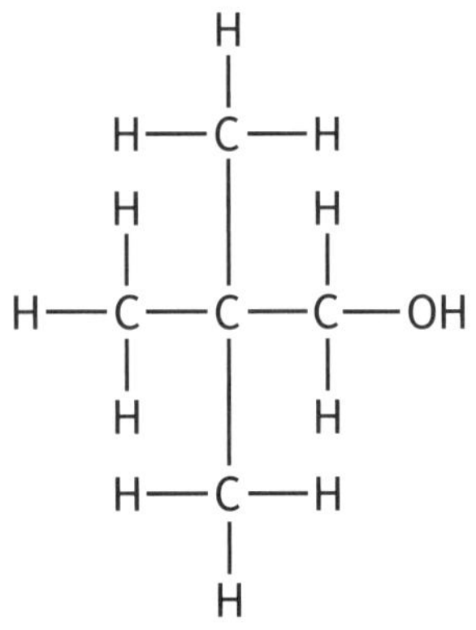

A   2,2,2-trimethylethanol

B   2,2-dimethylpropan-1-ol

C   2,2-dimethylpropan-3-ol

D   2,2-dimethylpentan-1-ol

8. Which of the following fatty acids is the most unsaturated?

A $C_{15}H_{29}COOH$

B $C_{15}H_{31}COOH$

C $C_{17}H_{31}COOH$

D $C_{17}H_{35}COOH$

9. Which of the following is **not** a step in a free radical chain reaction?

   A   Activation

   B   Initiation

   C   Propagation

   D   Termination

10. Which of the following is an isomer of ethyl propanoate ($CH_3CH_2COOCH_2CH_3$)?

   A   Methyl propanoate

   B   Pentan-2-one

   C   Pentanoic acid

   D   Pentane-1,2-diol

11. Essential oils are

   A   non-water soluble, non-volatile compounds

   B   non-water soluble, volatile compounds

   C   water soluble, non-volatile compounds

   D   water soluble, volatile compounds.

12. The enthalpy of combustion of a hydrocarbon is the enthalpy change when

   A   one mole of a hydrocarbon burns to give one mole of water

   B   one mole of a hydrocarbon burns to give one mole of carbon dioxide

   C   one mole of a hydrocarbon burns completely in oxygen

   D   one mole of a hydrocarbon burns in one mole of oxygen.

13. Which of the following is the strongest reducing agent?

   A   Fluorine

   B   Lithium

   C   Calcium

   D   Iodine

[Turn over

14.

$$TiCl_4 \quad + \quad 2Mg \quad \rightarrow \quad 2MgCl_2 \quad + \quad Ti$$

| mass of one mole = 189·9 g | mass of one mole = 24·3 g | mass of one mole = 95·3 g | mass of one mole = 47·9 g |

The atom economy for the production of titanium in the above equation is equal to

A $\dfrac{47·9}{189·9 + 24·3} \times 100$

B $\dfrac{47·9}{189·9 + (2 \times 24·3)} \times 100$

C $\dfrac{95·3 + 47·9}{189·9 + 24·3} \times 100$

D $\dfrac{(2 \times 47·9)}{189·9 + 24·3} \times 100$

15. The vitamin C content of a carton of orange juice was determined by four students. Each student carried out the experiment three times.

	Experiment 1 (mg/100 cm^3)	Experiment 2 (mg/100 cm^3)	Experiment 3 (mg/100 cm^3)
Student A	30·0	29·0	28·0
Student B	26·4	26·6	26·8
Student C	26·9	27·0	26·9
Student D	26·9	26·5	26·9

The most reproducible results were obtained by

A Student A

B Student B

C Student C

D Student D.

16. Cyanohydrin compounds can be made from carbonyl compounds by reacting the carbonyl compound with hydrogen cyanide (HCN).

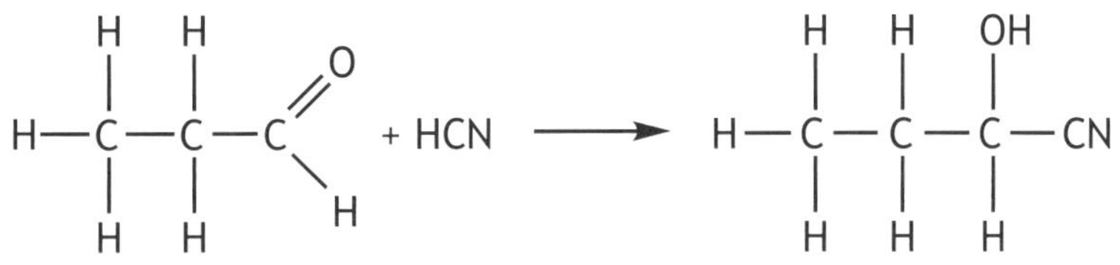

Which carbonyl compound would react with hydrogen cyanide (HCN) to form the following compound?

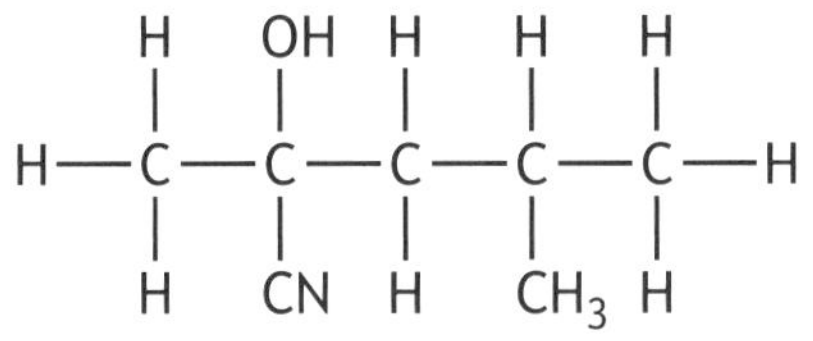

A

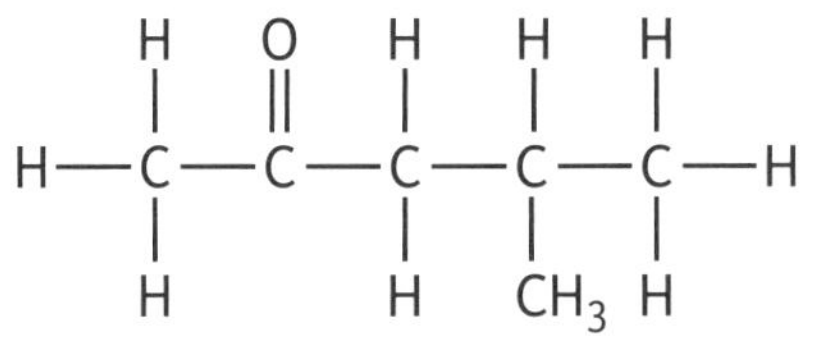

B

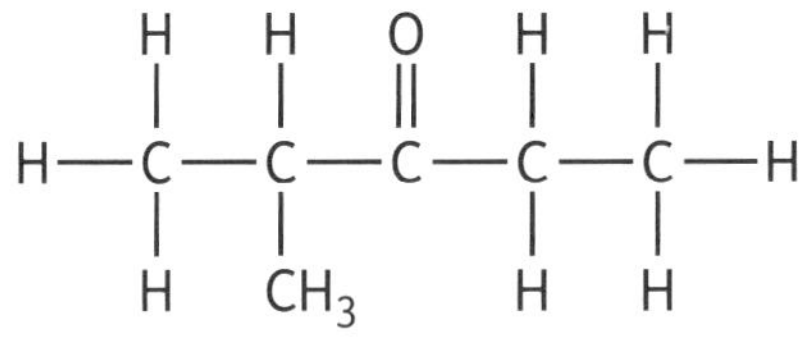

C

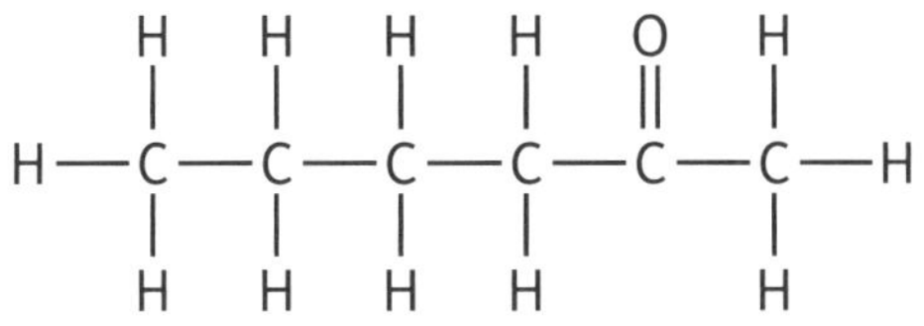

D

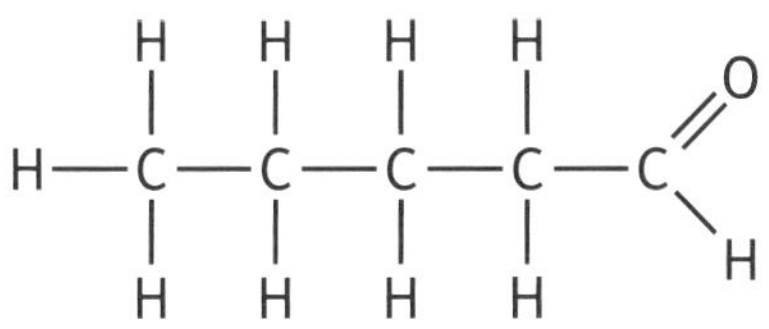

[Turn over

17. Chemical reactions are in a state of dynamic equilibrium only when

    A   the reaction involves no enthalpy change

    B   the concentrations of reactants and products are equal

    C   the activation energies of the forward and backward reactions are equal

    D   the rate of the forward reaction equals that of the backward reaction.

18. Bromine and hydrogen react together to form hydrogen bromide.

$$H_2(g) \;+\; Br_2(g) \;\longrightarrow\; 2HBr(g)$$

Bonds broken	Bonds made
H—H	2 × H—Br
Br—Br	

Bond	Bond enthalpy (kJ mol^{-1})
H—H	436
Br—Br	194
H—Br	366

The enthalpy change for this reaction, in kJ mol^{-1}, is

A   −102

B   +102

C   −264

D   +264.

19. Which of the following is a structural formula for glycerol?

A
$$
\begin{array}{l}
CH_2OH \\
| \\
CH_2 \\
| \\
CH_2OH
\end{array}
$$

B
$$
\begin{array}{l}
CH_2OH \\
| \\
CH_2OH
\end{array}
$$

C
$$
\begin{array}{l}
CH_2OH \\
| \\
CHOH \\
| \\
CH_2COOH
\end{array}
$$

D
$$
\begin{array}{l}
CH_2OH \\
| \\
CHOH \\
| \\
CH_2OH
\end{array}
$$

20. Which line in the table best describes the effect of adding a catalyst to the following reaction?

$$4NH_3(g) + 5O_2(g) \rightleftharpoons 4NO(g) + 6H_2O(g) \qquad \Delta H = -ve$$

	Position of equilibrium	Rate of forward reaction
A	unchanged	unchanged
B	unchanged	increased
C	moves to right	unchanged
D	moves to right	increased

[END OF SECTION 1. NOW ATTEMPT THE QUESTIONS IN SECTION 2
OF YOUR QUESTION AND ANSWER BOOKLET.]

[BLANK PAGE]

DO NOT WRITE ON THIS PAGE

H

National Qualifications 2017

Mark

X713/76/01

Chemistry
Section 1 — Answer Grid
and Section 2

MONDAY, 8 MAY

9:00 AM — 11:30 AM

Fill in these boxes and read what is printed below.

Full name of centre

Town

Forename(s)

Surname

Number of seat

Date of birth

Day   Month   Year   Scottish candidate number

Total marks — 100

SECTION 1 — 20 marks

Attempt ALL questions.

Instructions for the completion of Section 1 are given on *Page two*.

SECTION 2 — 80 marks

Attempt ALL questions.

You may refer to the Chemistry Data Booklet for Higher and Advanced Higher.

Write your answers clearly in the spaces provided in this booklet. Additional space for answers and rough work is provided at the end of this booklet. If you use this space you must clearly identify the question number you are attempting. Any rough work must be written in this booklet. You should score through your rough work when you have written your final copy.

Use **blue** or **black** ink.

Before leaving the examination room you must give this booklet to the Invigilator; if you do not, you may lose all the marks for this paper.

SECTION 1 — 20 marks

The questions for Section 1 are contained in the question paper X713/76/02.

Read these and record your answers on the answer grid on *Page three* opposite.

Use **blue** or **black** ink. Do NOT use gel pens or pencil.

1. The answer to each question is **either** A, B, C or D. Decide what your answer is, then fill in the appropriate bubble (see sample question below).

2. There is **only one correct** answer to each question.

3. Any rough working should be done on the additional space for answers and rough work at the end of this booklet.

Sample Question

To show that the ink in a ball-pen consists of a mixture of dyes, the method of separation would be:

    A   fractional distillation

    B   chromatography

    C   fractional crystallisation

    D   filtration.

The correct answer is **B** — chromatography. The answer **B** bubble has been clearly filled in (see below).

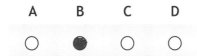

Changing an answer

If you decide to change your answer, cancel your first answer by putting a cross through it (see below) and fill in the answer you want. The answer below has been changed to **D**.

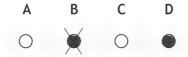

If you then decide to change back to an answer you have already scored out, put a tick (✓) to the **right** of the answer you want, as shown below:

SECTION 1 — Answer Grid

	A	B	C	D
1	○	○	○	○
2	○	○	○	○
3	○	○	○	○
4	○	○	○	○
5	○	○	○	○
6	○	○	○	○
7	○	○	○	○
8	○	○	○	○
9	○	○	○	○
10	○	○	○	○
11	○	○	○	○
12	○	○	○	○
13	○	○	○	○
14	○	○	○	○
15	○	○	○	○
16	○	○	○	○
17	○	○	○	○
18	○	○	○	○
19	○	○	○	○
20	○	○	○	○

[BLANK PAGE]

DO NOT WRITE ON THIS PAGE

[Turn over for next question

DO NOT WRITE ON THIS PAGE

MARKS | DO NOT WRITE IN THIS MARGIN

SECTION 2 — 80 marks

Attempt ALL questions

1. The elements sodium to argon make up the third period of the Periodic Table.

Na	Mg	Al	Si	P	S	Cl	Ar

(a) Name the element from the third period that exists as a covalent network.   1

(b) Ionisation energy changes across the period.

   (i) Explain why the first ionisation energy increases across the period.   1

   (ii) Write an equation, including state symbols, for the **second** ionisation energy of magnesium.   1

   (iii) The table shows the values for the first four ionisation energies of aluminium.

Ionisation energies (kJ mol^{-1})			
First	Second	Third	Fourth
578	1817	2745	11 577

Explain why there is a large difference between the third and fourth ionisation energies.   1

MARKS | DO NOT WRITE IN THIS MARGIN

1. **(continued)**

(c) The boiling point of chlorine is much higher than that of argon.

Explain **fully**, in terms of structure and the type of van der Waals forces present, why the boiling point of chlorine is higher than that of argon.

3

MARKS | DO NOT WRITE IN THIS MARGIN

2. Reactions involving iodine are commonly used to investigate rates of reaction.

   (a) One reaction involves hydrogen and iodine reacting together to form hydrogen iodide.

$$H_2(g) \ + \ I_2(g) \ \rightleftharpoons \ 2HI(g)$$

   (i) This reaction is thought to occur by initially breaking bonds in one of the reactants.

   Explain, using bond enthalpies, which bond is more likely to break first during this reaction.

   1

   (ii) The graph shows the distribution of kinetic energies of reactant molecules in the gas mixture at 300 °C.

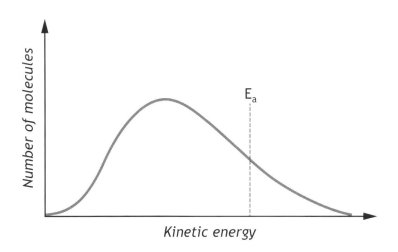

   Add a second curve to the graph to show the distribution of kinetic energies at 400 °C.

   1

   (An additional graph, if required, can be found on *Page thirty-five*)

MARKS | DO NOT WRITE IN THIS MARGIN

2. (a) (continued)

(iii) The reaction to produce hydrogen iodide is exothermic.

$$H_2(g) \ + \ I_2(g) \ \rightleftharpoons \ 2HI(g)$$

(A) State the effect of increasing temperature on the position of equilibrium.

1

(B) State why changing the pressure has no effect on this equilibrium reaction.

1

MARKS | DO NOT WRITE IN THIS MARGIN

2.   (a)   (continued)

(iv) The potential energy diagram for the reaction between hydrogen and iodine is shown.

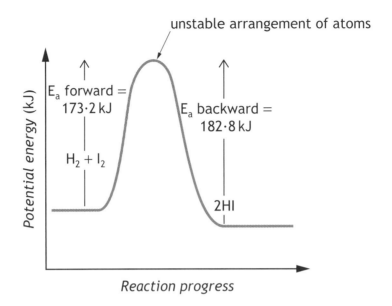

(A)   State the term for the unstable arrangement of atoms.   1

(B)   Calculate the enthalpy change, in kJ, for the forward reaction.   1

(C)   Platinum can be used as a catalyst for this reaction.

State the effect that platinum would have on the activation energy for the reaction.   1

2. (continued)

(b) The reaction between iodide ions, $I^-(aq)$, and persulfate ions, $S_2O_8^{2-}(aq)$, is used to investigate the effect of changing concentration on rate of reaction. The relative rate of the reaction is determined by mixing the reactants in a beaker and recording the time taken for the mixture to change colour.

The results of the investigation are shown in the table.

Experiment	Concentration of $I^-(aq)$ (mol l^{-1})	Concentration of $S_2O_8^{2-}(aq)$ (mol l^{-1})	Time (s)	Relative rate (s^{-1})
1	0·04	0·05	241	0·00415
2	0·06	0·05	180	0·00556
3	0·08	0·05		0·00819
4	0·1	0·05	103	0·00971

(i) The instructions state that a dry beaker must be used for each experiment.

Suggest a reason why the beaker should be dry.　1

(ii) Calculate the time, in seconds, for the reaction in experiment 3.　1

(iii) Explain why decreasing the concentration of iodide ions lowers the reaction rate.　1

MARKS | DO NOT WRITE IN THIS MARGIN

3. The leaves of the rhubarb plant are considered poisonous because they contain high levels of oxalic acid.

Oxalic acid is a white, water-soluble solid. It is a dicarboxylic acid that has the structural formula shown.

Oxalic acid reacts with bases to form salts.

It can also be oxidised by strong oxidising agents to form carbon dioxide gas. The oxidation equation for oxalic acid is shown.

$$H_2C_2O_4 \rightarrow 2CO_2 + 2e^- + 2H^+$$

Using your knowledge of chemistry, comment on how the mass of oxalic acid in a rhubarb leaf could be determined.

3

MARKS

DO NOT WRITE IN THIS MARGIN

4. Pentyl butanoate is responsible for some of the flavour in apricots and strawberries.

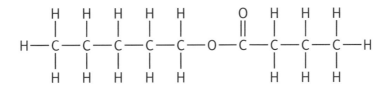

(a) Hydrolysis of pentyl butanoate using sodium hydroxide produces an alcohol and the salt of the carboxylic acid.

   (i) Name the alcohol that would be formed when pentyl butanoate is hydrolysed. 1

   (ii) Draw a structural formula for the sodium salt of the carboxylic acid that would be formed. 1

(b) Fats and oils belong to the same class of compounds as pentyl butanoate.

   (i) Name this class of compounds. 1

   (ii) When a fat is hydrolysed using sodium hydroxide, sodium salts of fatty acids are produced.

   State a use for sodium salts of fatty acids. 1

MARKS | DO NOT WRITE IN THIS MARGIN

4. **(b) (continued)**

(iii) Hydrolysis of fats using hydrochloric acid produces fatty acids. Stearic acid is a fatty acid that can be made from hydrolysis of beef fat. It is a fuel sometimes found in fireworks.

During combustion, stearic acid ($C_{17}H_{35}COOH$) produces 623 kJ of energy **per mole of CO_2 produced.**

$$C_{17}H_{35}COOH \ + \ 26O_2 \ \longrightarrow \ 18CO_2 \ + \ 18H_2O$$

mass of
one mole
$= 284$ g

Calculate the energy released, in kJ, by combustion of 10 grams of stearic acid.

2

5. Sulfur dioxide is a colourless, toxic gas that is soluble in water and more dense than air.

(a) One laboratory method for preparation of sulfur dioxide gas involves adding dilute hydrochloric acid to solid sodium sulfite. The sulfur dioxide gas produced is dried by bubbling the gas through concentrated sulfuric acid. The sulfur dioxide gas can then be collected.

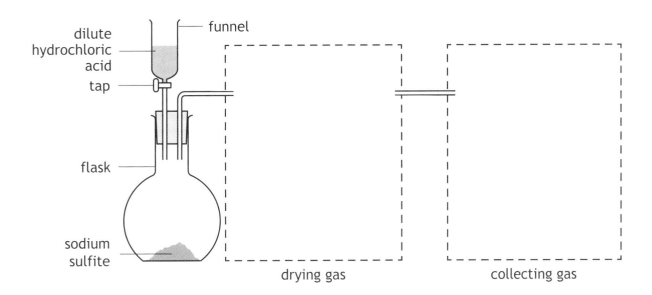

MARKS | DO NOT WRITE IN THIS MARGIN

(i) Complete the diagram by drawing:

in the first box, apparatus suitable for drying the sulfur dioxide gas;

in the second box, apparatus suitable for collecting the gas.   **2**

(An additional diagram, if required, can be found on *Page thirty-five*)

MARKS | DO NOT WRITE IN THIS MARGIN

5. **(a)** **(continued)**

(ii) 0·40 g of sodium sulfite, Na_2SO_3, is reacted with 50 cm³ of dilute hydrochloric acid, concentration 1·0 mol l⁻¹.

$$Na_2SO_3(s) + 2HCl(aq) \rightarrow H_2O(\ell) + 2NaCl(aq) + SO_2(g)$$

mass of
one mole
= 126·1 g

Show, by calculation, that sodium sulfite is the limiting reactant.   **2**

(b) Another reaction that produces sulfur dioxide gas involves combustion of carbon disulfide in the reaction shown.

$$CS_2(\ell) + 3O_2(g) \rightarrow CO_2(g) + 2SO_2(g)$$

Calculate the enthalpy change, in kJ mol⁻¹, for this reaction using the following information.   **2**

$$C(s) + O_2(g) \rightarrow CO_2(g) \qquad \Delta H = -393\cdot5\,kJ\,mol^{-1}$$
$$S(s) + O_2(g) \rightarrow SO_2(g) \qquad \Delta H = -296\cdot8\,kJ\,mol^{-1}$$
$$C(s) + 2S(s) \rightarrow CS_2(\ell) \qquad \Delta H = +87\cdot9\,kJ\,mol^{-1}$$

5. **(continued)**

MARKS | DO NOT WRITE IN THIS MARGIN

(c) The graph shows results for an experiment to determine the solubility of sulfur dioxide in water.

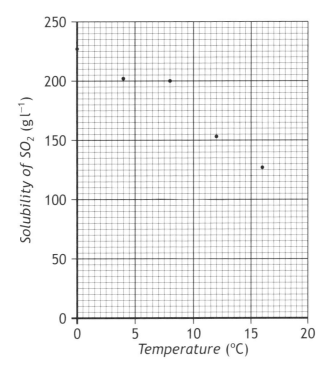

(i) Determine the solubility of sulfur dioxide, in gl^{-1}, in water at 10 °C.   **1**

(ii) Information about sulfur dioxide and carbon dioxide is shown in the table.

	Shape of molecule	Electronegativity difference between elements	Solubility in water at 25 °C $(g\,l^{-1})$
carbon dioxide	linear $O{=}C{=}O$	1·0	1·45
sulfur dioxide	bent $O{\overset{S}{\diagup}}{\diagdown}O$	1·0	94

Explain **fully** why carbon dioxide is much less soluble in water than sulfur dioxide is in water.   **2**

MARKS | DO NOT WRITE IN THIS MARGIN

6. A student was carrying out an investigation into alcohols, aldehydes and ketones.

(a) The student was given three alcohols labelled as **A**, **B** and **C**. These alcohols were all isomers with the formula C_4H_9OH.

    (i) Draw a structural formula for the secondary alcohol with the formula C_4H_9OH.

    1

    (ii) The student set up the following experiment.

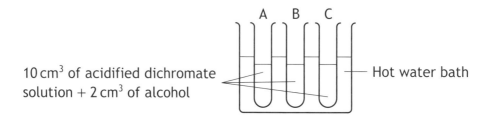

10 cm³ of acidified dichromate solution + 2 cm³ of alcohol   Hot water bath

Alcohol	Observation
A	Colour change
B	No change
C	Colour change

    (A) Suggest why a water bath is an appropriate method of heating the reaction mixture.

    1

    (B) Describe the colour change that would have been observed with alcohols **A** and **C**.

    1

    (C) Alcohol **B** is not oxidised.

        State the **type** of alcohol which cannot be oxidised by acidified dichromate solution.

    1

MARKS | DO NOT WRITE IN THIS MARGIN

6.   (a)   (continued)

(iii)   The student set up a second experiment with alcohol **A**.

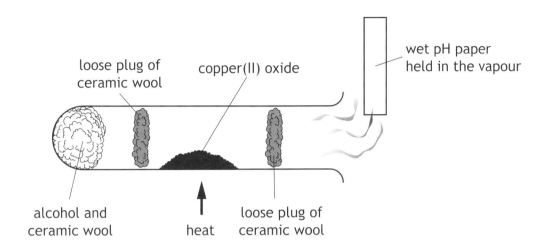

Hot copper(II) oxide is an oxidising agent.

(A)   When alcohol **A** (C_4H_9OH) is oxidised the product turns the pH paper red.

Suggest a name for the product.   1

(B)   Complete the ion-electron equation for the oxidation reaction.   1

$$C_4H_9OH \qquad \rightarrow \qquad C_4H_8O_2$$

MARKS | DO NOT WRITE IN THIS MARGIN

6. **(continued)**

(b) The student found the following information about the boiling points of some aldehydes.

Aldehyde	Molecular formula	Boiling point (°C)
	$C_5H_{10}O$	102
	$C_6H_{12}O$	130
	$C_7H_{14}O$	153
	$C_5H_{10}O$	95
	$C_5H_{10}O$	75
	$C_6H_{12}O$	119
	$C_6H_{12}O$	111

(i) Name the aldehyde that has a boiling point of 119 °C.   1

(ii) Predict the boiling point, in °C, of the following molecule.   1

MARKS | DO NOT WRITE IN THIS MARGIN

6. (b) (continued)

(iii) Using information from the table, describe one way in which differences in structure affect the boiling point of **isomeric** aldehydes. **1**

(iv) State what would be observed when an aldehyde is gently heated with Tollens' reagent. **1**

(c) Ketones contain a carbonyl group.

Name the type of intermolecular interaction between ketone molecules. **1**

MARKS | DO NOT WRITE IN THIS MARGIN

7. Some people take iron tablets as a dietary supplement. Iron tablets may contain iron(II) sulfate.

(a) A student was investigating the iron(II) content of iron tablets. A work card gave the following instructions for preparing an iron tablet solution.

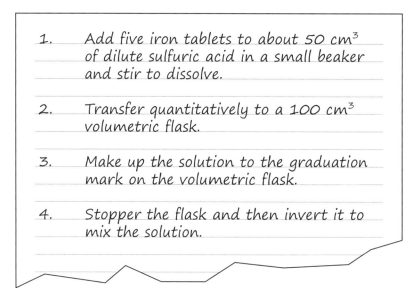

1. Add five iron tablets to about 50 cm³ of dilute sulfuric acid in a small beaker and stir to dissolve.

2. Transfer quantitatively to a 100 cm³ volumetric flask.

3. Make up the solution to the graduation mark on the volumetric flask.

4. Stopper the flask and then invert it to mix the solution.

To 'transfer quantitatively' means that **all** of the iron tablet solution must be transferred into the volumetric flask.

Describe how this is carried out in practice.   1

(b) The concentration of iron(II) ions (Fe^{2+}) in this iron tablet solution can be determined by a redox titration with permanganate (MnO_4^-) solution.

$$5Fe^{2+}(aq) + 8H^+(aq) + MnO_4^-(aq) \rightarrow 5Fe^{3+}(aq) + Mn^{2+}(aq) + 4H_2O(\ell)$$

(i) Suggest why it is **not** necessary to add an indicator to this titration.   1

MARKS | DO NOT WRITE IN THIS MARGIN

7. (b) (continued)

    (ii) Suggest why the titration must be carried out under acidic conditions.

        1

    (iii) Three $25.0 \, cm^3$ samples of the iron tablet solution were titrated with a standard solution of $0.020 \, mol \, l^{-1}$ permanganate (MnO_4^-). The results are shown below.

Sample	Volume of permanganate (cm^3)
1	14·9
2	14·5
3	14·6

    (A) State why the volume of permanganate used in the calculation was taken to be $14.55 \, cm^3$, although this is not the average of the three titres in the table.

        1

    (B) Calculate the concentration, in $mol \, l^{-1}$, of iron(II) ions in the iron tablet solution.

        3

$$5Fe^{2+}(aq) + 8H^+(aq) + MnO_4^-(aq) \rightarrow 5Fe^{3+}(aq) + Mn^{2+}(aq) + 4H_2O(\ell)$$

MARKS | DO NOT WRITE IN THIS MARGIN

7. (b) (iii) (continued)

(C) State what is meant by the term **standard solution**. 1

(D) Name an appropriate piece of apparatus which could be used to measure $25 \cdot 0\,cm^3$ samples of iron tablet solution. 1

(c) In a different experiment, five iron tablets were found to contain 0·00126 moles of iron(II) ions.

Calculate the average mass, in **mg**, of iron present in **one** tablet. 1

(d) It is recommended an adult female takes in 14·8 mg of iron per day.

100 g of a breakfast cereal contains 12·0 mg of iron.

Calculate the percentage of the recommended daily amount of iron provided for an adult female by a 30 g serving. 2

MARKS | DO NOT WRITE IN THIS MARGIN

8. Skin care products contain a mixture of polar covalent, non-polar covalent and ionic compounds. The compounds need to stay mixed within the product.

Skin care products also need to spread easily and remain on the skin for a period of time, as well as provide physical and chemical protection from the sun. In order to do this, skin care products contain a range of chemicals including water, fats and oils, antioxidants, minerals and sun block.

Using your knowledge of chemistry, explain the role of different chemicals in skin care products.

3

[BLANK PAGE]

DO NOT WRITE ON THIS PAGE

9. Dishwasher tablets contain chemicals which remove dirt from dishes.

(a) Dishwasher tablets include detergents. These molecules act like soaps to allow mixing of fat-soluble dirt and water.

(i) During the cleaning process, the detergent molecules combine with fat-soluble dirt.

A simplified diagram of a detergent molecule is shown.

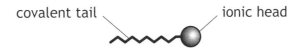

covalent tail   ionic head

Complete the diagram below to show how detergent molecules combine with fat-soluble dirt.   **1**

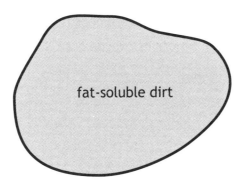

fat-soluble dirt

(An additional diagram, if required, can be found on *Page thirty-six*)

(ii) State the term used to describe the non-polar, hydrocarbon tail of a detergent molecule.   **1**

[**Turn over**

MARKS | DO NOT WRITE IN THIS MARGIN

9. **(continued)**

(b) Dishwasher tablets produce the bleach hydrogen peroxide, H_2O_2. One action of this oxidising agent is to oxidise food.

(i) Suggest another action of the bleach produced by the dishwasher tablets.

1

(ii) Hydrogen peroxide decomposes to form water and oxygen.

$$2H_2O_2(\ell) \quad \rightarrow \quad 2H_2O(\ell) \quad + \quad O_2(g)$$

A dishwasher tablet produces 0·051 g of hydrogen peroxide (mass of one mole = 34 g).

Calculate the volume of oxygen that would be produced when 0·051 g of hydrogen peroxide decomposes.

Take the volume of 1 mole of oxygen gas to be 24 litres.

3

(c) Enzymes are commonly added to dishwasher tablets. These are used to break down proteins.

(i) The proteins are broken down into small, water-soluble molecules.

Name the small, water-soluble molecules made when proteins are broken down completely.

1

9. (c) (continued)

(ii) The structure of a section of protein chain found in egg white is shown.

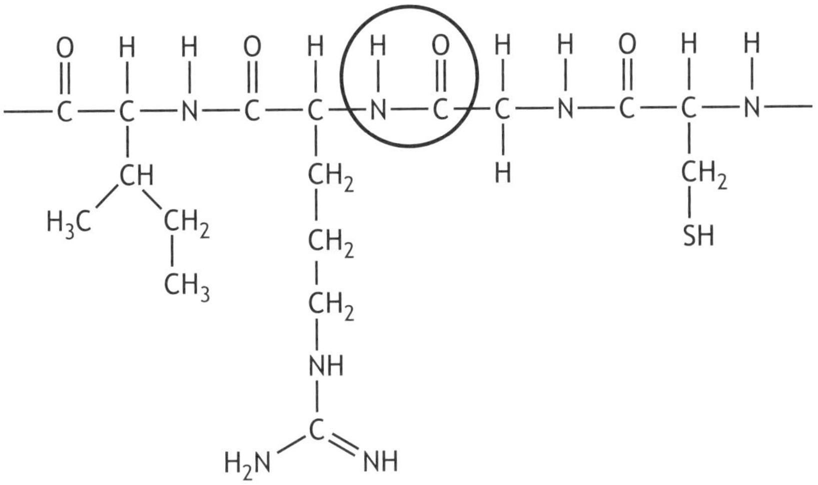

(A) Name the functional group circled.

1

(B) Draw a structural formula for **one** of the molecules that would be made when this section of egg white protein chain is completely broken down.

1

(iii) As part of the program in the dishwasher, the conditions in the dishwasher change so that the enzyme molecules no longer work because they change shape.

(A) State the term used to describe the change in shape of enzyme molecules.

1

(B) Suggest a change in conditions which would cause the enzyme molecules to change shape.

1

MARKS DO NOT WRITE IN THIS MARGIN

9. (continued)

(d) A bleach activator is frequently added to dishwasher tablets to speed up the bleaching reaction. One common bleach activator is TAED.

TAED could be produced in a process which involves a number of stages.

(i) The first stage in producing TAED is shown below.

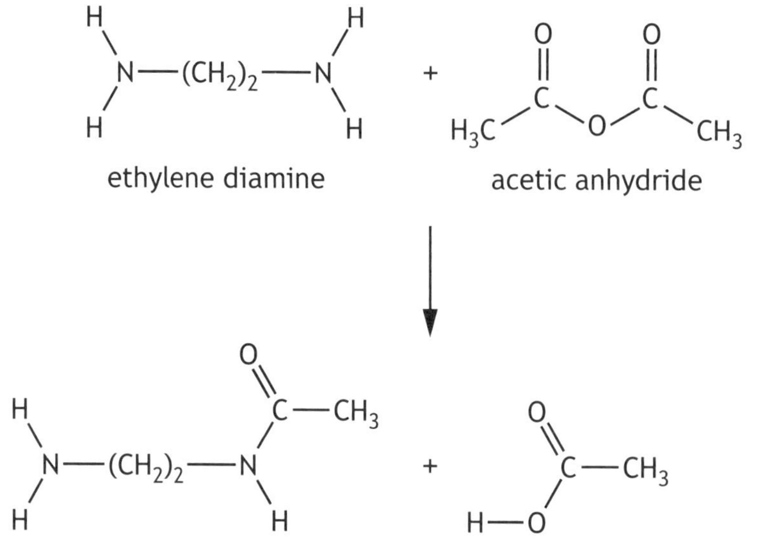
ethylene diamine
acetic anhydride

Suggest a name for this type of reaction. 1

9. (d) (continued)

(ii) The final stage in the process producing TAED is shown below.

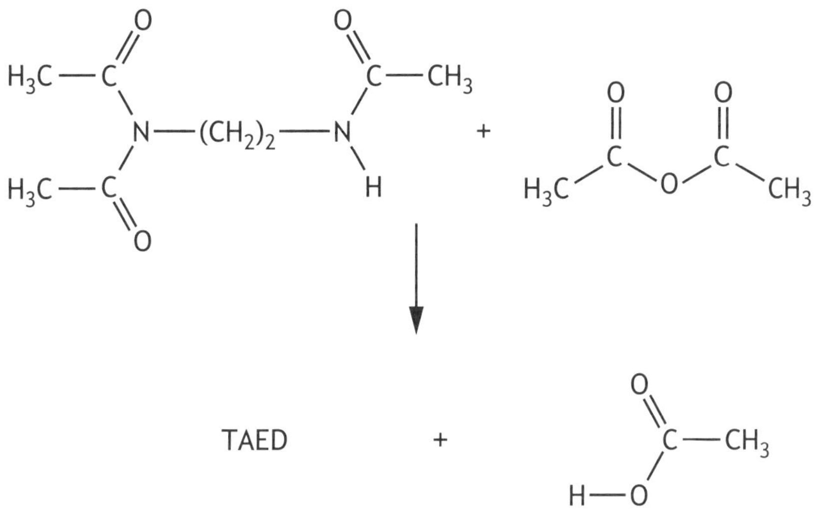

Draw a structural formula for TAED.   1

10. Essential oils from the lavender plant are used in aromatherapy.

MARKS

(a) Gas chromatography can be used to separate and identify the organic compounds in lavender oils.

Chromatogram 1 — Lavender oil **A**

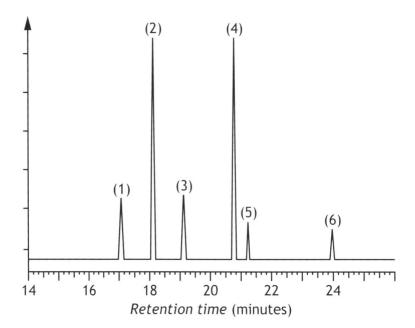

Peak	Component	Component peak area
1	1,8-cineole	7432
2	linalool	31 909
3	camphor	7518
4	linalyl acetate	27 504
5	geranyl acetate	3585
6	farnesene	1362

Total peak area   =   79 310

The relative concentration of each component can be calculated using the following formula.

$$\text{Relative concentration} = \frac{\text{Component peak area}}{\text{Total peak area}} \times 100 \ (\%)$$

(i) Calculate the relative concentration of linalool in lavender oil **A**.   1

MARKS | DO NOT WRITE IN THIS MARGIN

10. **(a)** **(continued)**

(ii) Different varieties of lavender oils have different compositions.

Chromatogram 2 — Lavender oil **B**

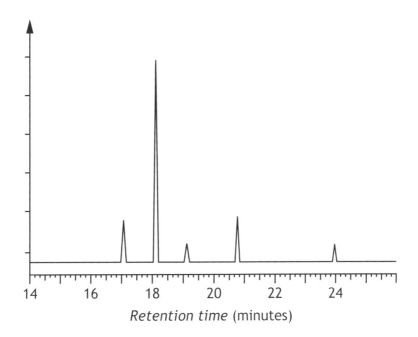

Retention time (minutes)

Identify the component found in lavender oil **A** that is missing from lavender oil **B**.

1

(b) A brand of mouthwash contains the component 1,8-cineole at a concentration of 0·92 mg per cm^3. The cost of 1 kg of 1,8-cineole is £59·10.

Calculate the cost, in pence, of 1,8-cineole that is present in a 500 cm^3 bottle of this brand of mouthwash.

2

MARKS | DO NOT WRITE IN THIS MARGIN

10. **(continued)**

(c) The component molecules found in lavender oils are terpenes or terpenoids.

(i) A chiral carbon is a carbon atom attached to **four** different atoms or groups of atoms.

An example is shown below.

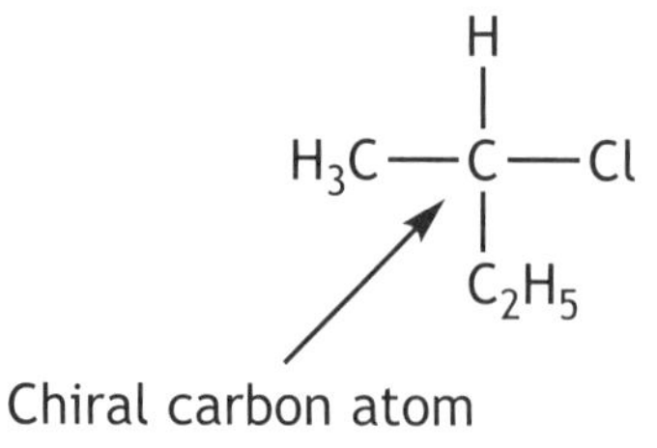

Chiral carbon atom

A molecule of the terpenoid linalool has a chiral carbon. Linalool has the following structure.

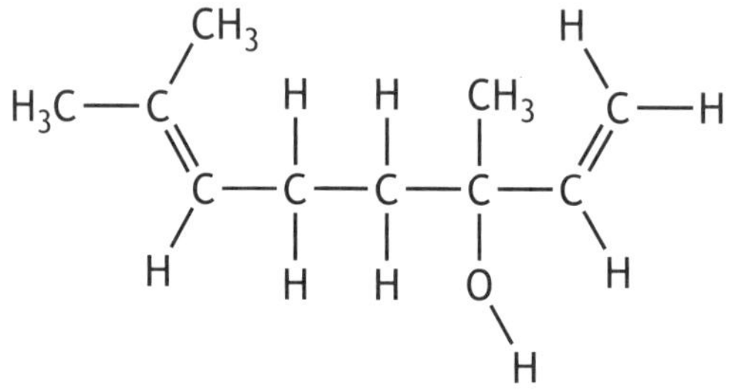

Circle the chiral carbon atom in the linalool structure. 1

(An additional diagram, if required, can be found on *Page thirty-six*)

(ii) Farnesene is a terpene consisting of **three** isoprene units (2-methylbuta-1,3-diene) joined together.

Write the molecular formula of farnesene. 1

[END OF QUESTION PAPER]

MARKS | DO NOT WRITE IN THIS MARGIN

ADDITIONAL SPACE FOR ANSWERS AND ROUGH WORK

ADDITIONAL DIAGRAM FOR USE IN QUESTION 2 (a) (ii)

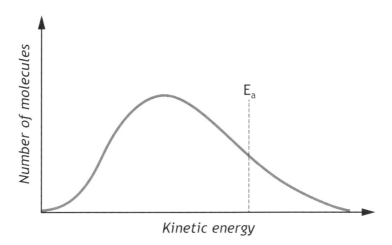

ADDITIONAL DIAGRAM FOR USE IN QUESTION 5 (a) (i)

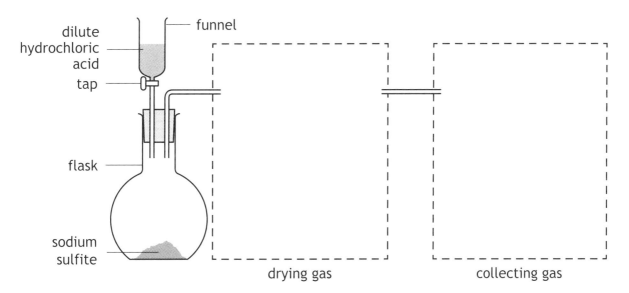

MARKS | DO NOT WRITE IN THIS MARGIN

ADDITIONAL SPACE FOR ANSWERS AND ROUGH WORK

ADDITIONAL DIAGRAM FOR USE IN QUESTION 9 (a) (i)

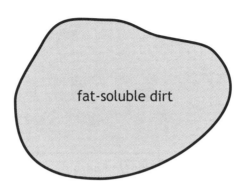

fat-soluble dirt

ADDITIONAL DIAGRAM FOR USE IN QUESTION 10 (c) (i)

ADDITIONAL SPACE FOR ANSWERS AND ROUGH WORK

MARKS DO NOT WRITE IN THIS MARGIN

ADDITIONAL SPACE FOR ANSWERS AND ROUGH WORK

HIGHER

Answers

HIGHER CHEMISTRY
2015

Section 1

Question	Response
1.	D
2.	C
3.	D
4.	A
5.	C
6.	B
7.	A
8.	B
9.	B
10.	A
11.	D
12.	D
13.	B
14.	C
15.	B
16.	A
17.	D
18.	A
19.	A
20.	C

Section 2

1. (a) Sulfur — London dispersion forces/van der Waals/intermolecular forces (1 mark)

    Silicon dioxide — covalent **or** polar covalent **or** covalent network bonds

    (1 mark)
    Maximum mark: 2

   (b) (i) Any structure for P_4S_3 that obeys valency rules

   Maximum mark: 1

   (ii) Sulfur has more protons in nucleus **or** sulfur has increased nuclear pull for electrons **or** increased nuclear charge

   Maximum mark: 1

   (iii) Correctly identify that the forces are stronger between sulfur (molecules) than between the phosphorus molecules (1 mark)

Correctly identifying that there are London dispersion forces between the molecules of both these elements (1 mark)

These forces are stronger due to sulfur structure being S8 whereas phosphorus is P4 (1 mark)
   Maximum mark: 3

2. (a) From graph, rate = 0·022
    t = 1/rate = 45s
   Maximum mark: 1

   (b) (i) Second line displaced to left of original. Peak of curve should be to the left of the original peak

   (ii) A vertical line drawn at a lower kinetic energy that the original Ea shown on graph
   Maximum mark: 1

3. (a) (i) Workable apparatus for passing steam through strawberry gum leaves. Steam must pass **through** the strawberry gum leaves (1 mark)

   Workable apparatus for condensing the steam and essential oil (1 mark)
   Maximum mark: 2

   (ii) (Fractional) distillation or chromatography
   Maximum mark: 1

   (b) (i) **1 mark** awarded for correct arithmetical calculation of moles of acid = 0·044 and moles alcohol = 0·063

   OR

   working out that 9·25g cinnamic acid would be needed to react with 2g methanol or 6·5g cinnamic acid would react with 1·41g methanol

   1 mark awarded for statement demonstrating understanding of limiting reactant.

   e.g. there are less moles of cinnamic acid therefore it is the limiting reactant

   OR

   0·0625 moles methanol would require 0·0625 moles cinnamic acid

   OR

   methanol is in excess therefore cinnamic acid is the limiting reactant.
   Maximum mark: 2

   (ii) (A) 52%

   1 mark is given for working out the theoretical yield ie 7·1g

   OR

   for working out both the moles of reactant used **AND** product formed ie both 0·044 moles and 0·023 moles

   1 mark is given for calculating the % yield, either using the actual and theoretical masses

   OR

   using the actual number of moles of products and actual number of moles of reactant
   Maximum mark: 2

(B) £24·59

1 mark for

Evidence for costing to produce of 3·7g
(£0·91)

OR

evidence of a calculated mass of cinnamic
acid × 14p

OR

evidence that 176g of cinnamic acid required

£12·80 would be using 100% yield

Maximum mark: 2

4. (a) Any one of the common compounds correctly
identified i.e.

citronellol/geraniol/anisyl alcohol

Maximum mark: 1

(b) Compounds that are common to the brand name
and counterfeit perfumes are present in lower
concentration in the counterfeit perfume

OR

Smaller volumes of compounds that are common to
the brand name and counterfeit perfumes are used
in the counterfeit perfume

Maximum mark: 1

(c) (i) Inert/will not react with the molecules (being
carried through the column)

Maximum mark: 1

(ii) Size (mass) of molecules / temperature of column.

Maximum mark: 1

(d) (i) Terpenes

Maximum mark: 1

(ii) (A) 3,7-dimethylocta-1,6-dien-3-ol

Maximum mark: 1

(B) Hydroxyl attached to C which is attached
to 3 other C atoms

OR

hydroxyl attached to a C that has no H
atoms attached

Maximum mark: 1

(e) 1·7 g (units not required)

5. This is an open ended question

1 mark: The student has demonstrated, at an appropriate
level, a limited understanding of the chemistry involved.
The student has made some statement(s) which is/are
relevant to the situation, showing that at least a little of
the chemistry within the problem is understood.

2 marks: The student has demonstrated a reasonable
understanding, at an appropriate level, of the chemistry
involved. The student makes some statement(s) which
is/are relevant to the situation, showing that the
problem is understood.

3 marks: The maximum available mark would be
awarded to a student who has demonstrated a good
understanding, at an appropriate level, of the chemistry
involved. The student shows a good comprehension
of the chemistry of the situation and has provided a
logically correct answer to the question posed. This
type of response might include a statement of the
principles involved, a relationship or an equation, and
the application of these to respond to the problem.

This does not mean the answer has to be what might be
termed an "excellent" answer or a "complete" one.

Points that could have been discussed include:
trends in Periodic properties such as covalent radius,
electronegativity and ionisation energies of the
elements in the groups; bonding and structures of the
elements in the groups.

Maximum mark: 3

6. (a) Heat breaks hydrogen bonds

Maximum mark: 1

(b) (i) Either of structures shown circled

Maximum mark: 1

(b) (ii) 50·5 ± 1 °C

Maximum mark: 1

(c) (i) Hydrolysis

Maximum mark: 1

(ii) (A) 5

Maximum mark: 1

(B) Correctly drawn amino acid structure

```
    H   H   O
    |   |   ||
H — N — C — C — OH
        |
        CH — CH₃
        |
        CH₃
```

```
    H   H   O
    |   |   ||
H — N — C — C — OH
        |
        CH₂
        |
        CH₂
        |
        C = O
        |
        OH
```

Maximum mark: 1

7. (a) 118/32

OR

3·69 mol CH3OH (1 mark)

3·69 × 24 = 88·5 litres (1 mark)

Maximum mark: 2

(b) (i) (A) Thermometer touching bottom or directly above flame

OR

temperature rise recorded would be greater than expected

Maximum mark: 1

(B) Distance between flame and beaker

OR

Height of wick in burner

Same type of beaker

Same draught proofing

Maximum mark: 1

(C) 2 concept marks and 1 arithmetic mark

Concept marks

Demonstration of the correct use of the relationship $E_h = cm\Delta T$ (1 mark) e.g. 4·18 × 0·1 × 23 OR 9·61

AND

Knowledge that enthalpy of combustion relates to 1 mol (1 mark) evidenced by scaling up of energy released

Correct arithmetic = –288 kJ mol⁻¹ (1 mark)

Maximum mark: 3

(ii) 0·799 (0·8)

Maximum mark: 1

(c) (i) If reactions are exothermic heat will need to be removed/If reactions are endothermic heat will need to be supplied

OR

Chemists can create conditions to maximise yield

Maximum mark: 1

(ii) Answer = +191 kJ mol⁻¹ (2)

Evidence of the use of all the correct bond enthalpies (1 mark) (412, 360, 463, 436, 743)

OR

Correct use of incorrect bond enthalpy values

Maximum mark: 2

8. (a) Calcium carbonate/carbon dioxide/ammonia/ calcium oxide all correctly identified in flow diagram (1 mark)

Ammonium chloride/sodium hydrogen carbonate/ sodium carbonate/water — all correctly identified in flow diagram (1 mark)

Maximum mark: 2

(b) (Adding brine) increases sodium ion concentration hence equilibrium shifts to right (1 mark)

Rate of forward reaction is increased (by addition of brine) (1 mark)

Maximum mark: 2

9. This is an open ended question

1 mark: The student has demonstrated, at an appropriate level, a limited understanding of the chemistry involved. The candidate has made some statement(s) at which is/ are relevant to the situation, showing that at least a little of the chemistry within the problem is understood.

2 marks: The student has demonstrated, at an appropriate level, a reasonable understanding of the chemistry involved. The student makes some statement(s) which is/are relevant to the situation, showing that the problem is understood.

3 marks: The maximum available mark would be awarded to a student who has demonstrated, at an appropriate level, a good understanding of the chemistry involved. The student shows a good comprehension of the chemistry of the situation and has provided a logically correct answer to the question posed. This type of response might include a statement of the principles involved, a relationship or an equation, and the application of these to respond to the problem. This does not mean the answer has to be what might be termed an "excellent" answer or a "complete" one.

Points that could have been discussed include: nature of crude oil and vegetable oil; cleaning action of detergents; emulsions.

Maximum mark: 3

10. (a) (i) 24 hours allows time for all of the zinc to react (1 mark)

No stopper allows hydrogen gas to escape from the flask (1 mark)

Maximum mark: 2

(ii) Zinc ions/impurities/metal ions/salts may be present in tap water

(b) (i) Pipette

Maximum mark: 1

(ii) 10 (Units not required, if given mg per litre, mg l⁻¹)

Maximum mark: 1

(c) Answer in range 4·6—4·8

(mg per litre, mg l⁻¹)

Maximum mark: 1

11. (a) Carboxyl/carboxylic (acid) group

Maximum mark: 1

(b) Esterification/condensation

Maximum mark: 1

(c)

Maximum mark: 1

(d) As molecular size (no. of carbon atoms) increases, the absorption decreases

Maximum mark: 1

12. (a) (i) 3 points

1 mark for rinsing the burette — rinse the burette with the thiosulfate/required solution/ with the solution to be put in it.

2 marks (1 mark each) for any 2 of the following points

- fill burette above the scale with thiosulfate solution
- filter funnel used should be removed
- tap opened/some solution drained to ensure no air bubbles
- (thiosulfate) solution run into scale
- reading should be made from bottom of meniscus

Maximum mark: 3

(ii) $2I^-(aq) \longrightarrow I_2(aq) + 2e^-$

Maximum mark: 1

(iii) 0·000062 (mol l^{-1})

Scheme of two "concept" marks, and one "arithmetic" mark

1 mark for knowledge of the relationship between moles, concentration and volume. This could be shown by one of the following steps:

Calculation of moles thiosulfate solution e.g. 0·001 × 0·0124 = 0·0000124

OR

calculation of concentration of iodine solution e.g. 0·0000062/0·1

OR

Insertion of correct pairings of values for concentration and volume in a valid titration formula

1 mark for knowledge of relationship between moles of thiosulfate and hypochlorite. This could be shown by one of the following steps:

Calculation of moles hypochlorite from moles thiosulfate — e.g. 0·0000124/2 = 0·0000062

OR

Insertion of correct stoichiometric values in a valid titration formula

1 mark is awarded for correct arithmetic through the calculation. This mark can only be awarded if both concept marks have been awarded.

Maximum mark: 3

(b) **1 mark** correct arithmetic, either 44·4 (litres) or 44400 (cm^3)

1 mark correct units

Maximum mark: 2

(c) (i) **1 mark** Ammonia is polar and trichloramine is non-polar

1 mark Explanation of this in terms of polarities of bonds **OR** electronegativity differences of atoms in bonds

Maximum mark: 2

(ii) Substances that have unpaired electrons

Maximum mark: 1

(iii) Propagation

Maximum mark: 1

13. (a) Aldehyde group correctly identified

Maximum mark: 1

(b) Ring form correctly drawn

Maximum mark: 1

HIGHER CHEMISTRY 2016

Section 1

Question	Response	Max Mark
1.	D	1
2.	A	1
3.	A	1
4.	B	1
5.	B	1
6.	A	1
7.	C	1
8.	D	1
9.	C	1
10.	B	1
11.	B	1
12.	C	1
13.	C	1
14.	B	1
15.	C	1
16.	A	1
17.	C	1
18.	D	1
19.	D	1
20.	A	1

Section 2

1. (a) The number of successful collisions will increase/ There will be a greater chance of successful collisions. **(1 mark)**

    OR

    More reactant particles will have energy equal to or greater than the activation energy. **(1 mark)**
    Maximum mark: 1

   (b) (i) 1 mark for drawing suitable experiment that will work **and** for indicating how **volume** will be measured eg collecting in a gas syringe or downward displacement of water from a measuring cylinder or similar.
    Maximum mark: 1

    (ii) 3·7 (volume strength) **Maximum mark: 1**

   (c) 8 (s) **Maximum mark: 1**

2. (a) (i) (The electron shells are pulled closer because) nuclear charge increases/the number of protons in the nucleus increases.
    Maximum mark: 1

    (ii) Two points are required.

    Understanding that the atom loses an electron (when the ion is formed). **(1 mark)**

    AND

    the sodium ion will only have two electron shells whereas the sodium atom has three electron shells.

    OR

    the sodium ion will have fewer electron shells (than the sodium atom). **(1 mark)**
    Maximum mark: 2

   (b) (i) As you go down the group the outer electron is more shielded from the nuclear pull **OR** less strongly attracted by the nucleus.
    Maximum mark: 1

    (ii) Second ionisation energy involves removal from an electron shell which is inner/full (whole)/ (more) stable/closer to the nucleus.

    OR

    Second electron is removed from an electron shell which is inner/full (whole)/(more) stable/ closer to the nucleus. **(1 mark)**

    The electron is less shielded from, or, more strongly attracted to the nucleus. **(1 mark)**
    Maximum mark: 2

   (c) (i) Any value in range 720–770 (kJ mol^{-1})
    Maximum mark: 1

    (ii) As the ionic radii (of the positive and/or the negative ion) increase, the lattice enthalpy decreases. **Maximum mark: 1**

3. (a) Covalent molecular **Maximum mark: 1**

   (b) n = m/gfm = 2900/58 = 50 **(1 mark)**

    V = n × Vm = 50 × 24 = 1200 (litres) **(1 mark)**

    Or by proportion

    58 g ⟶ 24 l **(1 mark)**

    2·9 kg ⟶ 24 × 2900/58ℓ

     = 1200 (ℓ) **(1 mark)**
    Maximum mark: 2

   (c) 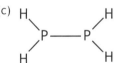 **Maximum mark: 1**

4. This is an open-ended question.

    1 mark: The student has demonstrated, at an appropriate level, a limited understanding of the chemistry involved. The student has made some statement(s) which is/are relevant to the situation, showing that at least a little of the chemistry within the problem is understood.

    2 marks: The student has demonstrated a reasonable understanding, at an appropriate level, of the chemistry involved. The student makes some statement(s) which is/are relevant to the situation, showing that the problem is understood.

    3 marks: The maximum available mark would be awarded to a student who has demonstrated a good understanding, at an appropriate level, of the chemistry involved. The student shows a good comprehension of the chemistry of the situation and has provided a logically correct answer to the question posed. This type of response might include a statement of the principles involved, a relationship or an equation, and the application of these to respond to the problem. This does not mean the answer has to be what might be termed an "excellent" answer or a "complete" one.
    Maximum mark: 3

5. (a) Glycerol **OR** glycerine **OR** propan(e)-1,2,3-triol

Maximum mark: 1

(b) (i) Polyunsaturated **Maximum mark: 1**

(ii) Octanoic acid **Maximum mark: 1**

(c) (i) Bromine solution is added to both until the bromine is no longer decolourised. (or reddish-brown colour remains). **(1 mark)**

More bromine would be required for the more unsaturated/olive oil.

OR

Less bromine would be required for the more saturated/coconut oil. **(1 mark)**

Maximum mark: 2

(ii) Hexane is a non-polar (solvent)/water is a polar (solvent). **Maximum mark: 1**

(iii) Coconut oil molecules can pack more closely together.

OR

Coconut oil has linear fatty acid chains/olive oil chains have bends/kinks (due to the double bonds). **(1 mark)**

There are stronger/more **intermolecular** forces between the molecules in coconut oil.

OR

The London dispersion forces (van der Waals' forces) between the molecules in coconut oil are stronger than in olive oils. **(1 mark)**

Maximum mark: 1

6. (a) Hexapeptide **Maximum mark: 1**

(b) Isoleucine-leucine-glycine-valine-serine

OR

Serine-valine-glycine-leucine-isoleucine

Maximum mark: 1

(c) Essential **Maximum mark: 1**

(d) (i) The peptide molecule:

must have contained an amino acid that is repeated in the sequence

OR

contained only four different amino acids (accept four different peptides).

OR

The peptide contains two amino acids:

with the same R$_f$ value

OR

that moved the same distance.

Maximum mark: 1

(ii)

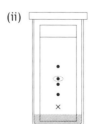

Maximum mark: 1

(e) (i)

Maximum mark: 1

(ii) 30 g — units required **(2 marks)**

Partial marks:

Correctly calculated mass of mushrooms without units **(1 mark)**

Appropriate units **(1 mark)**

Mass (1)	Units (1)
0·03	kg
30	g
30000	mg

Maximum mark: 2

7. (a) Any reason from list:

• UV light is damaging/harmful to skin.

• UV light causes sunburn.

• UV light can break bonds/molecules in skin.

• UV light damages collagen.

• UV light can cause skin cancer.

• UV light ages skin.

• UV light causes photo ageing.

• UV light creates free radicals/initiates free-radical chain reactions.

• Sunblocks contain free-radical scavengers.

Maximum mark: 1

(b) (i) Species (atoms/molecules, etc.) with unpaired electrons **Maximum mark: 1**

(ii) Initiation **Maximum mark: 1**

(iii) Carboxylic acid **OR** carboxyl group

Maximum mark: 1

(iv) The hydroxyl/functional group is attached to a carbon that is attached to two other carbons.

OR

The hydroxyl/functional group is attached to a carbon which has only one hydrogen attached.

Maximum mark: 1

(c) (i) A correct structural formula for butyl ethanoate

Maximum mark: 1

(ii) To condense any vapours or reactants/products which evaporate

OR

To act as a condenser **Maximum mark: 1**

(iii) 60·7/61%   **(2 marks)**

Calculates theoretical mass = 4·78g   **(1 mark)**

OR correctly calculates no of moles reactant
(0·054) and product (0·033)   **(1 mark)**

Calculating % yield using the actual and
theoretical masses, or using the actual number
of moles of products and actual number of
moles of reactant.   **Maximum mark: 2**

(iv) Condensation

OR

Esterification   **Maximum mark: 1**

8. (a) (i)

	Temperature (High/Low)	Pressure (High/Low)
Step 1	High	Low
Step 2	Low	High

Maximum mark: 2

(b) (i) Correct structure for 2-methylpropene
                              Maximum mark: 1

(ii) The proportion of the total mass of all starting
materials successfully converted into the
desired product is 100%.

OR

All the atoms in the reactants are converted
into the product you want./Mass of **product** is
equal to mass of reactants.

OR

No by-products/no waste products/only one
product is formed.   **Maximum mark: 1**

(c) 56 or +56   **(2 marks)**

Bond breaking

412 + 412 + 412 + 360 + 463 = 2059

Bond forming

743 + 412 + 412 + 436 = 2003

2059 − 2003 = 56 (kJ mol⁻¹)

A single mark is available if either of the following
operations is correctly executed.

Either

The five relevant values for the bond enthalpies
of the C-H, H-H, C-O, O-H and C=O bonds (or
multiples thereof) are retrieved from the data book
412,360,463,743,436 (ignore signs).

OR

If only four values are retrieved, the candidate
recognises that bond breaking is endothermic
and bond formation is exothermic and correctly
manipulates the bond enthalpy values they have
used to give their answer.   **Maximum mark: 2**

9. (a) (i) 4·93 (kJ) (no units required)

1 mark for extrapolating the graph and finding
the corrected temperature difference

$\Delta T = 11\cdot8$ (°C)

1 mark for use of the relationship $E_h = cm\Delta T$ to
calculate the $E_h = 4\cdot93$ (kJ)

Maximum of 1 mark for using $\Delta T = 11\cdot2$ which
gives $E_h = 4\cdot68$ (kJ)   **Maximum mark: 2**

(ii) To prevent (minimise) heat loss to the
surroundings

OR

Polystyrene cup is a poor conductor of heat/
insulator   **Maximum mark: 1**

(iii) 52·5 (kJ or kJmol⁻¹)   **Maximum mark: 1**

(b) −414 kJ mol⁻¹ (no units required)

Partial marks

Evidence of understanding of reversal of first
equation in order to achieve the target equation.
Reversal of both equations would be taken as
cancelling.

OR

Evidence of understanding of multiplying first
equation by 2 in order to achieve the target
equation.   **(1 mark)**
                          Maximum mark: 2

10. This is an open-ended question.

1 mark: The student has demonstrated, at an
appropriate level, a limited understanding of the
chemistry involved. The student has made some
statement(s) which is/are relevant to the situation,
showing that at least a little of the chemistry within the
problem is understood.

2 marks: The student has demonstrated a reasonable
understanding, at an appropriate level, of the chemistry
involved. The student makes some statement(s) which
is/are relevant to the situation, showing that the
problem is understood.

3 marks: The maximum available mark would be
awarded to a student who has demonstrated a good
understanding, at an appropriate level, of the chemistry
involved. The student shows a good comprehension
of the chemistry of the situation and has provided a
logically correct answer to the question posed. This
type of response might include a statement of the
principles involved, a relationship or an equation, and
the application of these to respond to the problem.
This does not mean the answer has to be what might
be termed an "excellent" answer or a "complete" one.
                          Maximum mark: 3

11. (a) (i) A drawing that shows a flask with a long narrow
neck and a single gradation mark which goes
completely across, or is labelled, on the narrow
neck   **Maximum mark: 1**

(ii) Accurate method for volume measurement e.g.
uses pipette   **(1 mark)**

Describes weighing by difference or using a
tared balance   **(1 mark)**
                          Maximum mark: 2

(iii) All points plotted correctly within tolerance of
½ a box   **(1 mark)**

Best fit line **(1 mark)**   **Maximum mark: 2**

(iv) (A) The dissolved gas/bubbles will affect the
density/mass/volume of solution
                          Maximum mark: 1

(B) 3·43%   **Maximum mark: 1**

(v) 34·98 (g)/35·0 (g)   **Maximum mark: 1**

(b) (i) Correctly drawn aldehyde group

$$\overset{O}{\underset{H}{\overset{\|}{-C}}}$$

Maximum mark: 1

(ii) $C_6H_{12}O_6 + H_2O \longrightarrow C_6H_{12}O_7 + 2H^+ + 2e$

Maximum mark: 1

(iii) Blue to orange/(brick)red/brown/yellow/green

Maximum mark: 1

(iv) 0·0099 (mol l^{-1}) **(3 marks)**

Partial marks

1 mark for knowledge of the relationship between moles, concentration and volume. This could be shown by any one of the following steps:

- Calculation of number of moles of Cu^{2+}

  e.g. 0·025 × 0·0198 = 0·000495

- Calculation of concentration of reducing sugars

  e.g. 0·0002475 ÷ 0·025

- Insertion of correct pairings of values of concentration and volume in titration formula

  e.g. $\dfrac{0·025 \times 19·8}{n_1} = \dfrac{C_{RS} \times 25·0}{n_2}$

1 mark for knowledge of relationship between Cu^{2+} and reducing sugars. This could be shown by any one of the following steps:

- Calculation of moles of reducing sugars from moles Cu^{2+}

  e.g. 0·000495 ÷ 2 = 0·0002475

- Insertion of correct stoichiometric values in titration formula

  e.g. $\dfrac{0·025 \times 19·8}{2} = \dfrac{C_{RS} \times 25·0}{1}$

1 mark is awarded for correct arithmetic throughout the calculation. Maximum mark: 3

12. (a) (i) Hydrogen bonding Maximum mark: 1

(ii) (More) branching lowers the boiling point.

(1 mark)

The shorter the alcohol, the lower the boiling point./The longer the carboxylic acid the lower the boiling point.

OR

The nearer the ester link is to the right-hand side (of the molecule), the higher the boiling point. **(1 mark)**

Maximum mark: 2

(iii) Any temperature between 99 and 124 (°C)

(data value 116°C) Maximum mark: 1

(b) (i) Peaks labelled in order

2,3,1,4 Maximum mark: 1

(ii) 5 Maximum mark: 1

HIGHER CHEMISTRY 2017

Section 1

Question	Answer	Max Mark
1.	A	1
2.	D	1
3.	C	1
4.	B	1
5.	D	1
6.	C	1
7.	B	1
8.	C	1
9.	A	1
10.	C	1
11.	B	1
12.	C	1
13.	B	1
14.	B	1
15.	C	1
16.	A	1
17.	D	1
18.	A	1
19.	D	1
20.	B	1

Section 2

1. (a) Silicon **(1 mark)**

(b) (i) Increasing/greater/stronger/higher nuclear charge (holds electron more tightly)

OR

Increasing number of protons **(1 mark)**

(ii) $Mg^+(g) \longrightarrow Mg^{2+}(g) + e^-$ **(1 mark)**

(iii) Fourth ionisation energy involves removal of an electron from an electron shell which is inner/full (whole)/(more) stable/closer to the nucleus

OR

fourth electron is removed from an electron shell which is inner/full (whole)/(more) stable/ closer to the nucleus.

OR

removing third electron is taking from an outer/a part full shell

OR

taking an electron from a full shell requires more energy (than removing from a part full shell)

OR

taking an electron from a part full shell requires less energy (than removing from a full shell)

OR

fourth electron is less shielded than the third electron

OR

third electron is more shielded than the fourth electron **(1 mark)**

(c) **1 mark:** Correctly identify that there are stronger/more (Van der Waals) forces between chlorine (molecules) than between the argon (atoms)

1 mark: Correctly identifying that the van der Waals forces present in both these elements are London dispersion forces

1 mark: Chlorine molecules (Cl_2) have more electrons than argon atoms (Ar). **Maximum mark: 3**

2. (a) (i) I-I bond is weaker/has a lower bond enthalpy value (so will break more easily)

OR

I_2 (151 kJ mol^{-1}) is less than H_2 (436 kJ mol^{-1}), (so will break more easily). **(1 mark)**

(ii) Peak of curve should be further to the right and no higher than the original line. **(1 mark)**

(iii) (A) Equilibrium will shift to the reactant side/left (hand side). **(1 mark)**

(B) There are the same/equal volume/number of moles/molecules (of gases) on each side (of the equation).

OR

Pressures of reactants and products are equal. **(1 mark)**

(iv) (A) Activated complex **(1 mark)**

(B) −9·6 (kJ) If candidate has calculated from graph values

OR

−9 (kJ) If candidate has calculated using bond enthalpies

Answer must include the negative sign **(1 mark)**

(C) Decrease/lower it **(1 mark)**

(b) (i) To keep the concentration (of the reactants) constant.

OR

Adding water will change/affect/dilute/decrease the concentration (of the reactants)

OR

To keep the total volume constant. **(1 mark)**

(ii) 122·1 (accept 122) (s) **(1 mark)**

(iii) The number of (successful) collisions will decrease.

OR

Less chance of (successful) collisions

OR

The frequency of (successful) collisions will decrease. **(1 mark)**

3. This is an open ended question

1 mark: The student has demonstrated, at an appropriate level, a limited understanding of the chemistry involved. The candidate has made some statement(s) at which is/are relevant to the situation, showing that at least a little of the chemistry within the problem is understood.

2 marks: The student has demonstrated, at an appropriate level, a reasonable understanding of the chemistry involved. The student makes some statement(s) which is/are relevant to the situation, showing that the problem is understood.

3 marks: The maximum available mark would be awarded to a student who has demonstrated, at an appropriate level, a good understanding of the chemistry involved. The student shows a good comprehension of the chemistry of the situation and has provided a logically correct answer to the question posed. This type of response might include a statement of the principles involved, a relationship or an equation, and the application of these to respond to the problem. This does not mean the answer has to be what might be termed an 'excellent' answer or a 'complete' one. **Maximum mark: 3**

4. (a) (i) Pentan-1-ol **(1 mark)**

(ii)

(1 mark)

(b) (i) ester **(1 mark)**

(ii) soap **(1 mark)**

(iii) 395 or −395 (kJ) **(2 marks)**

Partial marking — one mark can be awarded for:

the correct application of number of moles of stearic acid

eg

10/284 × 623 or 0·0352 × 623

(=21·94)   (=21·93)

or

10/284 × 18 or 0·0352 × 18

(=0·634)   (=0·634)

OR

the correct application of the stoichiometry

eg

the energy change for 1 mole of stearic acid as 623 × 18 = 11214 (kJ)

or

284 g ⟵⟶ 623 × 18 **Maximum mark: 2**

5. (a) (i) Diagram shows a workable method of bubbling through concentrated sulfuric acid. **(1 mark)**

Diagram for appropriate gas collection method i.e. using a gas syringe or upward displacement of air. **(1 mark)**

Maximum mark: 2

(ii) Calculating that 0·05 moles HCl would require 0·025 moles sodium sulfite and there are only 0·00317 moles of sodium sulfite

OR

Calculating that 0·00317 moles of sodium sulfite would require 0·00634 moles of HCl and there are 0·05 moles of HCl

Partial marking:

1 mark awarded for correct arithmetical calculation of moles of Na_2SO_3 (= 0·00317 mol) **AND** HCl = 0·05 mol)

OR

Calculating that 3·15 g sodium sulfite would be needed to react with 50 cm^3 hydrochloric acid and there are only 0·4 g of sodium sulfite

Partial marking:

1 mark awarded for correct arithmetical calculation of moles of acid (0·05) and correct application of stoichiometry to either reactant.

OR

Calculating that 6·3 cm^3 of (1 M) HCl would be needed to react with 0·4 g of sodium sulfite and there are 50 cm^3 (1M) HCl

Partial marking:

1 mark awarded for correct arithmetical calculation of moles of sodium sulfite (0·00317) and correct application of stoichiometry to either reactant. **(2 marks)**

(b) –1075 (kJ mol^{-1}) **(2 marks)**

Partial marking – treat as two concepts either would be acceptable for 1 mark

Evidence of understanding of reversal for third equation only in order to achieve the target equation.

Reversal of additional equations would be taken as cancelling

OR

evidence of understanding of multiplying for second equation by 2 in order to achieve the target equation. Multiplication of additional equations would be taken as cancelling. **(2 marks)**

(c) (i) 163 – 167 inclusive (g l^{-1}) **(1 mark)**

(ii) **1 mark** for carbon dioxide is non-polar due to its shape/dipoles cancelling out **and** sulfur dioxide is polar due to its shape/dipoles not cancelling out

1 mark for an explanation which links polarity of CO_2 and SO_2 molecules to the polarity of water **Maximum marks: 2**

6. (a) (i)

(1 mark)

(ii) (A) Reactants or products are flammable/could catch fire. **(1 mark)**

(B) orange to green/blue-green/blue **(1 mark)**

(C) Tertiary **(1 mark)**

(iii) (A) Butanoic acid

OR

(2-)methylpropanoic acid **(1 mark)**

(B) $C_4H_9OH + H_2O \longrightarrow C_4H_8O_2 + 4H^+ + 4e^-$ **(1 mark)**

(b) (i) 2-methylpentanal **(1 mark)**

(ii) Any temperature between 166 and 181 ($^\circ$C) **(1 mark)**

(iii) (More) branching lowers the boiling point (of isomeric aldehydes). **(1 mark)**

(iv) Silver mirror/silver precipitate **(1 mark)**

(c) (Permanent) dipole to (permanent) dipole **(1 mark)**

7. (a) Rinse beaker and transfer the rinsings/washings to the flask **(1 mark)**

(b) (i) The reaction is self-indicating.

OR

Potassium permanganate can act as its own indicator.

OR

Reaction changes colour. **(1 mark)**

(ii) To provide H^+ ions for the reaction. **(1 mark)**

(iii) (A) **1 mark** for any of the following

• first titre is a rough (or approximate) result/practice

• first titre is not accurate/not reliable/ rogue

• first titre is too far away from the others

• you take average of concordant/close results

(B) 0·0582 (mol l^{-1}) **(3 marks)**

Partial marks can be awarded using a scheme of two "concept" marks.

1 mark for knowledge of the relationship between moles, concentration and volume. This could be shown by any one of the following steps:

• Calculation of moles MnO_4^- solution eg 0·02 × 0·01455 = 0·000291

• Calculation of concentration of Fe^{2+} solution eg 0·001455/0·025

• Insertion of correct pairings of values for concentration and volume in a valid titration formula eg

$$\frac{0 \cdot 02 \times 14 \cdot 55}{n_1} = \frac{C_{Fe^{2+}} \times 25 \cdot 0}{n_2}$$

1 mark for knowledge of relationship between moles of MnO_4^- and Fe^{2+}. This could be shown by one of the following steps:

• Calculation of moles Fe^{2+} from moles MnO_4^- – eg 0·000291 × 5 = 0·001455

• Insertion of correct stoichiometric values in a valid titration formula eg

$$\frac{0 \cdot 02 \times 14 \cdot 55}{1} = \frac{C_{Fe^{2+}} \times 25 \cdot 0}{5}$$

Maximum marks: 3

(C) A solution of accurately/exactly/precisely
known concentration **(1 mark)**

(D) Pipette **(1 mark)**

(c) 14 mg, 14·06 mg, 0·01406 g **(1 mark)**

(d) 24%, 24·3% **(2 marks)**

Partial marks

1 mark awarded for

30 g would contain 3·6 mg

1 mark for

$$\frac{\text{any calculated mass}}{14\cdot8} \times 100 \qquad \textbf{Maximum marks: 2}$$

8. This is an open ended question

1 mark: The student has demonstrated, at an appropriate level, a limited understanding of the chemistry involved. The candidate has made some statement(s) at which is/are relevant to the situation, showing that at least a little of the chemistry within the problem is understood.

2 marks: The student has demonstrated, at an appropriate level, a reasonable understanding of the chemistry involved. The student makes some statement(s) which is/are relevant to the situation, showing that the problem is understood.

3 marks: The maximum available mark would be awarded to a student who has demonstrated, at an appropriate level, a good understanding of the chemistry involved. The student shows a good comprehension of the chemistry of the situation and has provided a logically correct answer to the question posed. This type of response might include a statement of the principles involved, a relationship or an equation, and the application of these to respond to the problem. This does not mean the answer has to be what might be termed an 'excellent' answer or a 'complete' one.

Maximum marks: 3

9. (a) (i) A drawing similar to

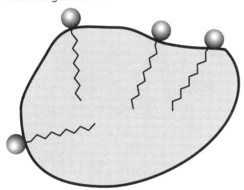

Diagram shows **at least one** detergent molecule. All tails shown should be **fully** inside the fat-soluble dirt. **(1 mark)**

(ii) hydrophobic **(1 mark)**

(b) (i) To break down coloured compounds/removes stains/kill bacteria/kill fungi/inactivate viruses/germs **(1 mark)**

(ii) 18 cm³/0·018 litres **with correct unit (3 marks)**

Partial marking — 1 mark can be awarded for two of the three steps shown below correctly calculated:

1. number of moles of H_2O_2

2. mole ratio applied

3. calculated number of moles of O_2 multiplied by 24 (24000)

If processed by proportion

68 g ⟷ 24 l (24000 cm³)

1 mark

OR

0·051 g ⟷ 0·036 l (36 cm³)

1 mark

1 mark for correct units. **Maximum marks: 3**

(c) (i) amino acids **(1 mark)**

(ii) (A) Amide/amide link/peptide link **(1 mark)**

(B) Any of the shown amino acids:

$$HO-\overset{\overset{\displaystyle O}{\|}}{C}-HC-NH_2$$

with $CH(H_3C)(CH_2-H_3C)$ side chain

$$HO-\overset{\overset{\displaystyle O}{\|}}{C}-CH-NH_2$$

with $-CH_2-CH_2-CH_2-NH-C(NH_2)(=NH)$ side chain

$$HO-\overset{\overset{\displaystyle O}{\|}}{C}-CH_2-NH_2$$

$$HO-\overset{\overset{\displaystyle O}{\|}}{C}-HC-NH_2$$

with $-CH_2-SH$ side chain

(1 mark)

(iii) (A) Denaturing **(1 mark)**

(B) Temperature increase/pH **(1 mark)**

(d) (i) Condensation   **(1 mark)**

(ii)

(1 mark)

10. (a) (i) 40·23/40·2/40 (%)   **(1 mark)**

(ii) geranyl acetate/peak 5   **(1 mark)**

(b) 2·7p/3p

Partial marking — 1 mark can be awarded for:

Evidence of scaling up to 500 cm^3

eg 460 mg of 1,8-cineole

OR

Evidence for determining a correct cost for a calculated mass of 1,8-cineole

eg 0·92 mg costs 0·00544 pence   **Maximum marks: 2**

(c) (i)

(1 mark)

(ii) $C_{15}H_{24}$ OR $(C_5H_8)_3$ **(1 mark)**

Acknowledgements

Permission has been sought from all relevant copyright holders and Hodder Gibson is grateful for the use of the following:

An extract from 'Patterns in the Periodic Table' © Royal Society of Chemistry (2015 Section 2 page 14);

Image © sima/Shutterstock.com (2016 Section 2 page 20).